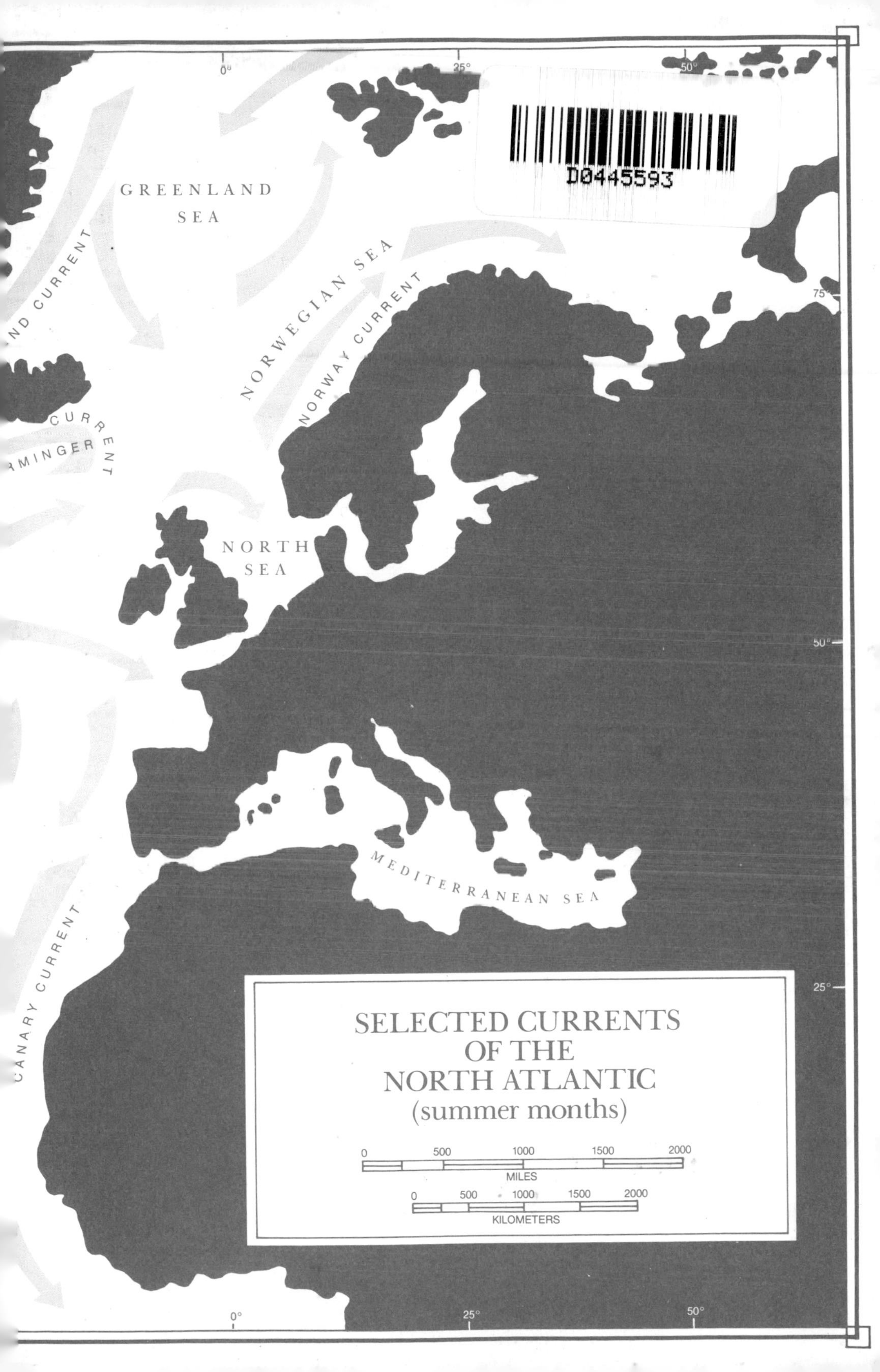
GREENLAND SEA
NORWEGIAN SEA
NORWAY CURRENT
NORTH SEA
MEDITERRANEAN SEA
CANARY CURRENT
SELECTED CURRENTS OF THE NORTH ATLANTIC (summer months)
MILES
KILOMETERS
0°
25°
50°
75°
D0445593

THE GULF STREAM

Books by William H. MacLeish

OIL AND WATER

THE GULF STREAM

551.
471
M

THE GULF STREAM

ENCOUNTERS WITH THE BLUE GOD

William H. MacLeish

ILLUSTRATIONS BY
SARAH LANDRY

Houghton Mifflin Company
Boston · 1989

R/V OCEANUS LIBRARY

Woods Hole Oceanographic Institution
Purchase Order No. 21198
24 Feb. 1989

Copyright © 1989 by William H. MacLeish

ALL RIGHTS RESERVED

For information about permission to reproduce selections from this book, write to Permissions, Houghton Mifflin Company, 2 Park Street, Boston, Massachusetts 02108.

Library of Congress Cataloging-in-Publication Data

MacLeish, William H., date.
The Gulf Stream: encounters with the blue god.
Bibliography: p.
Includes index.
1. Gulf Stream. I. Title.
GC296.M33 1989 551.47'1 88-8317
ISBN 0-395-40621-8

Printed in the United States of America

S 10 9 8 7 6 5 4 3 2 1

A portion of the chapter "Following On" appeared in slightly different form in *Motorboat* magazine. An adaptation of the book appeared in *Smithsonian* magazine.

Book design by Robert Overholtzer

Endpaper map by George Ward

For

Ishbel MacLeish Campbell

who has given me many things,
including this book

Acknowledgments

THE IDEA OF SOMEONE riding the Gulf Stream, hitchhiking around its gyre, tickled the fancy of a lot of people who are out there for a living — fishing those waters, steaming through them, sinking instruments in them, diving in them, overflying them. The help they gave me was essential to the writing of this book.

Many of those people appear in the book, and I thank them for their willingness to do that. Some spent hours and days reading chunks of early copy. Tom Rossby of the University of Rhode Island went over the entire manuscript, and so did Dana Densmore of Woods Hole and Alexander Campbell of Geneva, New York. Stanley Wilson of NASA checked my assertions about satellite oceanography. William Fowler of Northeastern University reviewed the historical material. Carl Wunsch of MIT looked at some of the oceanography. They, of course, are not responsible for the errors that may remain, but their work greatly increased my accuracy.

Some of the most generous people are not mentioned in these pages. Don Moser, editor of *Smithsonian* magazine, supported and encouraged the project from the beginning. Jane Cushman, my agent, kept me from the dwindles. Others took the time to lead me through theory and practice totally foreign to my experience. I think of William von Arx, Joseph Pedlosky, Nick Fofonoff, Terrence Joyce, and William Simmons, of Woods Hole; Cyrus Gordon of New York; James McCarthy of Harvard, and

Francis Bretherton of the National Center for Atmospheric Research; and John Casey and Merton Ingham of the National Marine Fisheries Service laboratory in Kingston, Rhode Island. John Walter of the Library of Congress and the editors of the Franklin papers at Yale led me to some satisfying discoveries.

Other writers have been drawn to the Gulf Stream, and I would like to pay tribute to two of them I knew. F. G. Walton Smith, together with Henry Chapin, wrote a book called *The Ocean River,* published in 1952 by Charles Scribner's Sons. I mention the book here rather than in the bibliography because copies are difficult to find. Thomas Lineaweaver of Woods Hole was writing about the Stream when he died. His book would have been a fine one.

I have been working on a documentary film based on the book, and the experience — the contrary pressures of writing and filming — has been on the whole beneficial, I think, to both enterprises. My thanks to John Borden, the producer, and Joana Hattery, the associate producer, and to the staff of Peace River Films, for insights, for new looks at old thoughts.

Robie Macauley, my editor, gave me a sense of where I was in the book without recourse to coordinates. Chris Jerome, the copy editor, showed me how to reshape and tighten what I had written in ways I still find marvelous. And through these three years and more, through great difficulties, my wife, Elizabeth Libbey, gave me, with her poetry and her presence, the best of all these generous gifts.

William H. MacLeish
Charlemont, Massachusetts
June 23, 1988

Contents

THE GULF STREAM

1

Introduction to a Gyre

I AM TAKEN by rivers.

I am used to small ones — the South, flowing through the Massachusetts hill town where I grew up; the Deerfield, running clear and proper below the hayfields of the hill town where I live now.

My father, who was a fine student of the Berkshires, once said that a river was "stone's refusal of water." I have gone by that, and by the smells of the water around here. Rust in the early spring, and the musk of ferment before the wind comes around and blows autumn in. The ice of November, scouting the banks and anchoring in the eddies and, some still, gelid night, closing in on the current. The rivers of January turn cataract blind, and there is nothing in them or along them that won't break before it bends.

I have not spent much of my adult life here. I am a wanderer. The longest time in one place was the decade I spent running a magazine of marine science at the Woods Hole Oceanographic Institution, under the shoulder of Cape Cod. I began thinking about this book shortly after I moved there. I had read what just about anyone interested in marine studies has read — the famous quote from Lieutenant Matthew F. Maury, an energetic naval administrator who developed quite a following in the mid-nineteenth century writing about what he called the physical geography of the sea. "There is a river in the ocean," Maury wrote. "In the severest droughts it never fails, and in the mightiest floods it never overflows. Its banks and its bottom are of cold water, while its current is of warm. The Gulf of Mexico is its fountain and its mouth is in the Arctic Seas. It is the Gulf Stream."

The idea of a great salt river flowing through the Atlantic a hundred miles and more beyond the Cape astounded and fascinated me. I read about the river's history, the theories of its biology, its physics. I rode the river in motor ships and sailing vessels. I dove in it, fished along its edge, overflew it. I talked about it with oceanographers along the Atlantic coast of the United States and in Europe, with fishermen and sailors and Coast Guardsmen. I learned what I could, sporting with the myths, struggling — an English major adrift — with the codes of fluid dynamics. I learned that the river is the most powerful current in the global ocean. That it is a prime agent in moderating climate. That it deeply influenced the invasion of the Americas by Western culture. That it is the most spectacular part of a great swirl in the North Atlantic, a gyre — the North Atlantic subtropical gyre. And that it is not a river.

Rivers — the Deerfield, the South — flow downhill. Part of the Gulf Stream system does, too: the level of the current drops by a hand's span along the Florida peninsula. But basically, the Stream as a whole flows *around* its hill — the slightly mounded Sargasso Sea. It is the slave not of gravity and the friction of banks and bottom, as is the Deerfield, but of gravity balanced against forces arising from the rotation of the earth. To one degree or another its flow, like other motions in ocean and atmosphere, approaches what fluid dynamicists call geostrophic circulation.

Henry Stommel, a dean of American oceanography, describes the Stream as "a narrow ribbon of high-velocity water acting as a boundary that prevents the warm water of the Sargasso Sea . . . from overflowing the colder, denser waters on the inshore side." No, not a river. And yet we all, oceanographers included, call it the Stream. We speak of meanders, eddies. We do not wish to correct the fluvial metaphor. Which is but one of many reasons why science prefers the certainty of numbers to the seduction of words. Stommel thinks that "a scientist is on pretty safe ground as long as he stays close to the mathematical formalism." Still, he says: "You remember the old prayers for those in peril on the sea? Perhaps there should be a special prayer for those in peril of trying to explain the sea." Amen.

And yet I must use what is familiar to me, the exercise of fantasy, to describe what is most unfamiliar to me, the forces that shape the Stream and most other oceanic movement. To take the metes and bounds of those waters, we must stand back. We must hitch a ride on a rocket, sit in a spaceship. Then we can see the oceans as they are, a domination of blue under the ermine scarves of the air.

We sit in our high station, and the globe moves below. Motion is the music of this sphere, motion along its sun track, motion around its axis, motion of the axis itself. Hundreds of miles above the Amazon basin, we wait for the Andes to come under. Forests, peaks, stationary to all who live among them, roll eastward toward the night. They appear to creep, yet they tear along, there on the equator, at more than a thousand miles an hour. They appear to go end over end in their equatorial circuit, while at the poles, the ice spins in place.

Our instruments tell us that the poles are not perpendicular to the plane of the earth's passage around the sun. From that asymmetry come the seasons. During our flight, the Northern Hemisphere is tilted sunward, into summer. We can also sense that insolation is far more intense directly below than at the high latitudes angling away from us over the curve of the sphere. The forces arising from this inequality, coupled with those of the earth's gravitational field, coupled with the earth's motion, are the Aeolus and Poseidon of fluid dynamics.

These forces are apt to be comparatively gentle. Some, in a sense, aren't there at all. They are concepts invented to enable scientists to set aside, with some safety, the more irritating complexities of the planet's dynamical equilibrium. One such invention is named after a nineteenth-century French civil engineer,

Gaspard Gustave de Coriolis, a motion man so enamored of his work that he once wrote a paper applying his mathematical principles to billiards. There is no such thing as a Coriolis force actually turning moving bodies to the right in the Northern Hemisphere and to the left in the Southern Hemisphere. If an obliging rocketeer in Quito, immediately below us, were to touch one off right now, aimed at the North Pole, we in our satellite would see the missile streak in a straight line to its target. But to someone on the ground looking down the flight path and tracking the rocket, the thing would seem to curve to the right.

The reason lies in rotation. Quito may be moving eastward at 1,000 miles an hour, but New Orleans, at 30 degrees north latitude, is only going 925, and Reykjavik, Iceland, at 64 degrees, is traveling less than 500. All are accelerating eastward at a rate that brings them full circle in twenty-four hours. Yet the farther north or south you go, the smaller the circle needs to be to do the job. So in terms of the earth's rotation, an object on the earth in high latitudes will move more slowly toward the east than will an object in low latitudes. That suborbital rocket launched northward at Quito will continue to move eastward at Quito's acceleration as it passes over locations moving progressively slower, so it will appear to the earthling to move to its right.

The Coriolis force acts on motion relative to the earth's motion: the higher the velocity, the stronger — more accurately, the less weak — the force. Even an object moving due east or west is affected by the earth's twist — except on the equator, where we see from our observation posts that there is no twist. Water particles at rest will tend to go eastward, each in its appropriate latitudinal circle. Each particle will, to the eye of the nearby beholder, who is also in his latitudinal transit, remain where it is, in equilibrium. But accelerate that particle eastward faster than the earth is moving and, in the Northern Hemisphere, the particle will turn to the right, toward the equator. Accelerate it westward so that it is moving slower than the earth and it will again bear right, toward the poles.

If we move in our minds from space to an office wall chart, we can see the structures such curvatures produce. They are the gyres of the world ocean and, like all ellipses, they have neither beginning nor end. Sailors and scientists have given them starts and stops by segmenting them and naming the segments. As the endpapers of this book show, the Gulf Stream off Florida is, to those oceanographers who observe the accepted nomenclature, not the Stream but the Florida Current. The Gulf Stream itself begins off Cape Hatteras and runs east and a bit north, out to seaward of the Grand Banks of Newfoundland.

Beyond the banks, in the Atlantic's interior, the Stream turns uncertain. Its recirculations appear to increase in great loops, most of them at depth, that rejoin the parent current far upstream. Some water brought north by the Stream moves eastward and northeastward, cooling as it goes. One movement, the North Atlantic Current, bears off northeast from the gyre, branching here and there but eventually pushing up along the Norwegian coast and over the top into the Barents Sea. Another carries the gyre in a bend toward the southeast and the Azores, and down toward the Canary Current off Africa. The Canary, in turn, bends west to help form the North Equatorial Current, the belt of warming water that runs under the trade winds. Some of that flow follows on through the teeth of the Antilles and into the Caribbean to close the gyre, to complete the Gulf Stream system.

The scales of the sea are such that it is extremely difficult to keep them in mind, let alone in perspective. Motion, shifts in properties, occur simultaneously across centimeters and across thousands of kilometers. At the global scale, the horizontal dimensions of the oceans dominate all else. Seawater is the seventy percent solution, covering almost three quarters of the earth. We are, as the first astronauts told us, living on a misnomer. At that same global scale, vertical dimensions are insignificant. The North Atlantic averages about twelve thousand feet in depth, and in its trenches plunges much farther. Yet if it were reduced in width and breadth to the size of this book, its depth would correspond roughly to the thickness of this page.

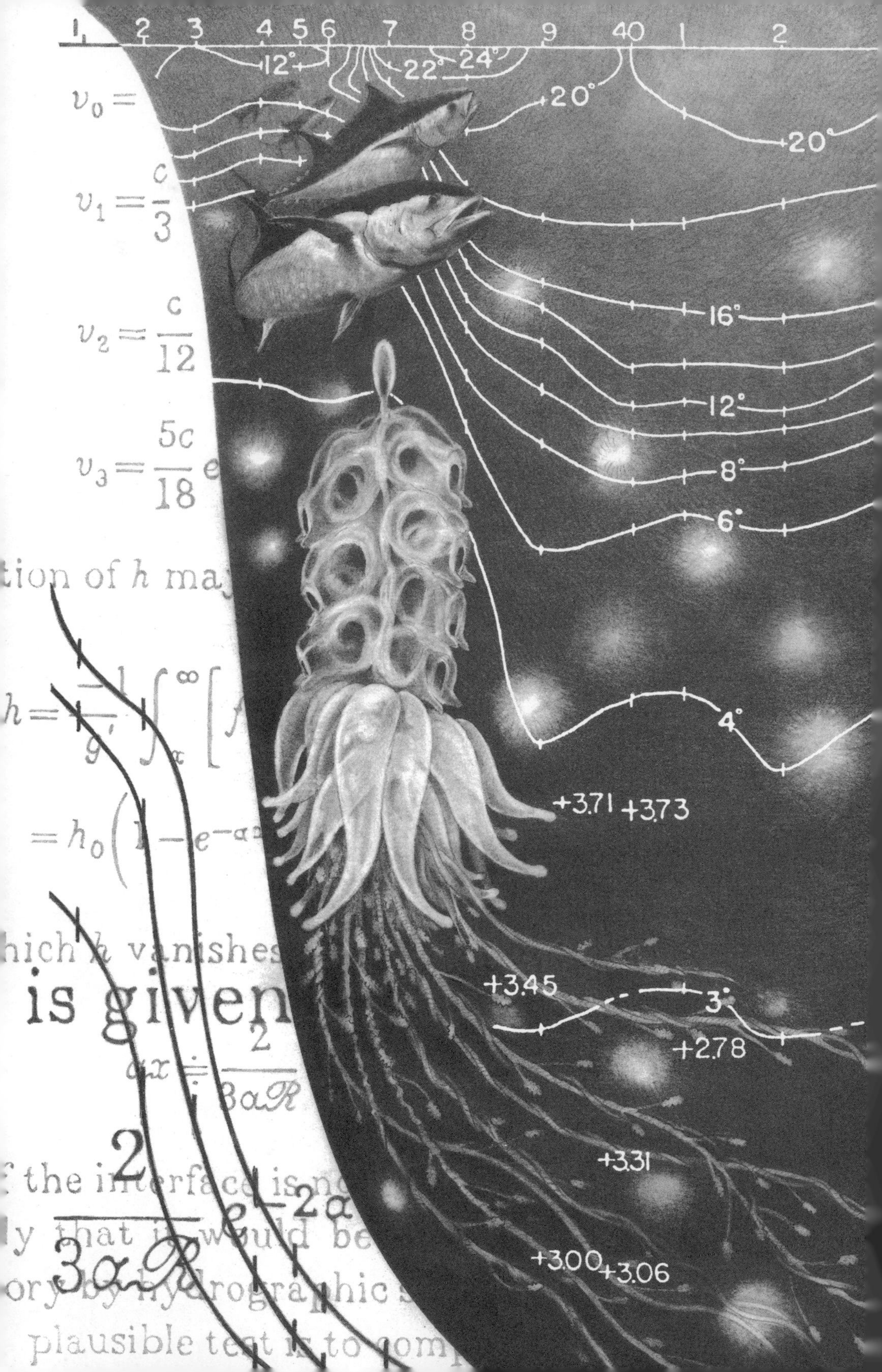

1 2 3 4 5 6 7 8 9 40 1 2

12° 22° 24° 20° 20°

$v_0 =$

$v_1 = \frac{c}{3}$

$v_2 = \frac{c}{12}$

$v_3 = \frac{5c}{18} e$

16° 12° 8° 6°

tion of h ma

$h = \frac{-1}{g'} \int_{a}^{\infty} \Big[f$

$= h_0 \Big(1 - e^{-a}$

4°

+3.71 +3.73

hich h vanishes

is given

$ax \div \frac{2}{3a\mathcal{R}}$

+3.45 3° +2.78

$\frac{2}{3a\mathcal{R}} e^{-2a}$

f the interface is n

y that it would be

ory by hydrographic s

plausible test is to comp

+3.31

+3.00 +3.06

I find it extraordinary that raiments so thin serve the wearer so well. Together with the atmosphere — and they are always together, two fluids at play in the spin of the earth — the oceans moisturize and moderate. Their currents, the Gulf Stream prime among them, disperse heat from the tropics toward the poles in such a way that temperatures over a broad band of latitudes remain within the fairly narrow range in which life flourishes.

In my reading and voyaging over the past three years, the scale that has interested me is the basin scale, the dimensions of the Gulf Stream's home water. A short while ago I came home from the North Atlantic on that fine edge between exhilaration and exhaustion. I lay in our meadow in the afternoon, looking out at the hills across the Deerfield River. They moved. They became the swells I had just left. I still had my sea legs; perhaps I also had my sea sight. I heard the wind flowing across the slope, bellying down in the hay. Thoughts kited up before it. I am safe home, in my element, breathing it, hearing it. Air is in me, around me. It is my life. Even in the heat haze, I could see the bottom of the bowl where the river winds, and the slopes of hardwoods and pine. I could walk to its far side before nightfall, always in my element.

Not so in the bowl I had just left, the great basin of the North Atlantic. No one has ever seen it whole. Only those uncomfortable few peering from their submersibles have ever seen parts of it. A guarantor of life in air, the ocean is supremely hostile to observation by those who live in air. Knowledge must come by proxy, from instruments of special capabilities; the ocean is opaque to the electromagnetic radiation we landlings take for granted, opaque to the light we live by.

With the lens of imagination, I tried to see what I could not. I conjured up the Atlantic basin, snaking across the equator, its artificial dividing line. In the north its deeps pinch out in large fjords, but at shallower depths it connects directly with the Arctic Ocean. With that addition, the Atlantic becomes the longest of oceans, a mighty strait from pole to pole — and beyond. There are few truly oceanic islands in the strait, most of them volcanic cones, like Iceland and the Azores, and most of those associated with the Mid-Atlantic Ridge, part of the great global welt where magma rises and solidifies into fresh sea floor.

I tried to find an image for the North Atlantic waters, a mere tenth or so of the volume of the world ocean, a mere 150 million cubic kilometers of alien fluid. That water is roughly nine hundred times denser than the air above our meadow. Its pressures at depth are shocking. When I started work at the Oceanographic Institution in Woods Hole, its director was fond of impressing audiences by holding up two Styrofoam representations of the old-time straw boater hat. One was about a man's medium. The other had been carried down to a couple of thousand meters in the basket of *Alvin*, the Oceanographic's submersible. Still perfectly formed, it wouldn't fit a chipmunk.

The stoves of the earth were at work in the meadow. I could feel them beneath my back. All the fine day, air had been rising over the hills and flowing away, its going in balance with the wind coming in low over the grass — a far more efficient heat engine than the oceanic version. The warm air above the Deerfield rises, mixing high. The sea takes its heat from the atmosphere above it. The wind acts to mix surface waters, but only to a depth of one hundred or two hundred meters. Those warmer waters float in a lens. Below lies the bulk of the sea, dark and cold. (Deep mixing does occur, but mainly in the polar seas, where water chills so much that its density carries it down to the abyss.)

The ocean below the mixed layer is normally stratified in what, for want of a more precise term, oceanographers call density layers. The density of a water parcel, controlled by its temperature, salinity, and the pressure acting on it, determines what layer it will seek. The planes of density surfaces often depart from the horizontal, usually at gentle slopes, and those inclinations are essential to ocean circulation.

At the sea surface, winds are the prime agents of horizontal movements. They drag across the sea as they drag through hay, and they tease the water into moving not only up and down, as waves do — and hay — but forward in a wind drift. In the North

Atlantic, atmospheric forcing of this sort produces the North Equatorial Current below the trade winds and the North Atlantic Drift under the westerlies.

Because of the rotation of the earth we saw from the spaceship, the sea moves to the right of the wind in the Northern Hemisphere and to the left in the south. Even well below the windblown surface, the water continues to move more and more to the right until, at about a hundred meters down, it barely creeps. The inertia of so much water flowing in so many patterns is such, one oceanographic textbook asserts, that if the sea winds were to drop in eternal doldrum, the waters they energized would remain in motion for some six months.

Thunder burbled far to the west of our meadow, and the hot wind strengthened. I thought, in my Atlantic reverie, of the warm Sargasso, the sea of weed where the bodies of men and ships were once thought to float for an eternity, suspended between surface and deeps. The winds work on the Sargasso, driving warm waters toward its center. The waters pile up, and a combination of forces establishes an imperceptible bulge in the western part of the sea, perhaps two meters high, sloping for hundreds of kilometers across the surface. Vanishingly small changes in dynamic topography can produce extraordinary changes in oceanic behavior. That Sargasso slope, that nothing, is enough to trigger processes that lead to the formation of the Gulf Stream.

Pressure under the center of the Sargasso lens is greater than that at the edge, so water at the center moves outward, toward areas of less pressure. As it does, it too comes under the sway of the Coriolis force, and turns right. The Stream, the high-velocity ribbon Henry Stommel talked about, forms the seaward side of a boundary between different water masses. On the landward side lie coastal waters that are cooler, fresher, far richer in life than those in either the Stream or the Sargasso.

The boundary — mariners call it variously the west wall or the north wall — lives up to its name on oceanographic charts of temperature and salinity gradients, which exaggerate the vertical to clarify what is going on. There, lines connecting, say, points of equal temperature appear to plunge, whereas in reality they may decline at a slope of one foot in a hundred. As they move toward the Sargasso, across the Stream, they return to near horizontal.

Were it not for that wall, palms might grow on Cape Hatteras and coral might encrust the wrecks on its shoals. But the wall endures. The warm waters circling the Sargasso and crowding into the Caribbean and the Gulf of Mexico under the trade winds cannot seriously breach it, though there are intrusions, forays, toward the coast. Instead, they deflect to north and east in Stommel's ribbon. There are other such boundary currents. The North Pacific gyre has one. It too carries the darkest of ocean colors, and its name, Kuroshio, means "black water" in Japanese. It has its own explicative theories, its own myths. But it cannot match the Stream.

Nor, in my bias, can the myths of other boundary currents match those of the Stream. The river in the sea is American, courtesy of Matthew Maury. The Europeans have their beliefs, their certainties. The Gulf Stream, they say, keeps northern Europe from the frigidities of Labrador, which lies at the same high latitude as Britain and Germany. The Gulf Stream, they say, accounts for the palms in southwestern Ireland; for the strange exotics in the gardens of the Isles of Scilly; for the fact that some northern European ports stay open in winter while others farther to the south surrender to the ice. Never mind that the relationship between the Stream and the flows heading east across the Atlantic is poorly understood at best. The myths win out. Hans Leip, a writer in Switzerland, invoked those myths thirty years ago while describing the beauties of spring around Lake Constance. "The moist . . . air," he claimed, "is warmed by that mysterious current which comes to Europe from the cauldrons of the West Indies."

On both sides of the North Atlantic, the notion has grown of the Gulf Stream as the engine of civilization. Climate makes the man. The Gulf Stream makes Gulf Stream Man, energetic, industrious. Leip again: "The radiation of [the Gulf Stream's] warm, moist breath, together with a certain irregularity in its supply of heat, produces that peculiar irritability of the Europeans, their restless and creative fever, and the varied development of European cultures." Over here, the Gulf Stream has been held responsible by believers for the plantation culture of the South, the cod fisheries of the North, and a good deal in between. Its name is in our songs: from the redwood forests to the Gulf Stream waters. In our commerce: a Gulf Stream raw bar caters to a college crowd just a few miles from my hill town. Even in our geography. There is a Gulf Stream in western Connecticut, a couple of hours south of the Deerfield. I can jump across it.

Moving water, I thought as I got up and walked through the hay toward the house, seems everywhere to stimulate the braggarts of culture. This, from *Uncle Tom's Cabin*, about the Mississippi:

> What other river of the world bears on its bosom to the ocean the wealth and enterprise of such another country? . . . Those turbid waters, hurrying, foaming, tearing along, an apt resemblance of that headlong tide of business which is poured along its waves by a race more vehement and energetic than any the world ever saw.

What other river? The Stream, of course. The mysterious river in the sea.

A round of thunder exploded just down the valley as I reached the kitchen door, and I went inside to watch the rain coming on.

2

The Blue God

I SAW THE MISSISSIPPI for the first time in the spring of 1986. I had made out the river before, from the impossible altitudes of transcontinental flights, but that was map reading. This was seeing. I stood in hot and weighted air on the bridge wing of the chemical carrier *Exxon Wilmington,* tied up just above Baton Rouge, Louisiana, and I thought: This is not stone's refusal of water, my father's vision of a river. This is water's acceptance of earth. We were supposed to be having an El Niño year, a time when ocean and atmosphere shift gears the slightest bit and bring death to billions of anchovies off Peru, storms to California, drought to the southeastern United States. There had been no high water to speak of that spring in the Mississippi, but below me the river ran with its stolen silts so fast it jittered and sucked along the black hull, and out in the full current, midway to the low green bank a half mile distant, trees came pushing along with their roots reaching out like rams.

Setting words to water is so hard; water does not stop for words. The Chippewas, who knew the river before the wing dams and levees, called it *Mee-zee-see-bee,* the father of waters. William Faulkner, the white writer, called it simply the Old Man.

The words that suited me best that hot May day looking out at the river, that riot, were written by a poet who grew up on the Mississippi and left it for England. "I do not know much about gods," T. S. Eliot wrote in *The Dry Salvages*; "but I think that the river/ Is a strong brown god — sullen, untamed and intractable . . . " And, he wrote, "The river is within us . . . "

Baton Rouge is the upper point of navigation for big ships. From there up, it's mostly tows, assemblages of muscular boats and cargo barges that can take up acres of the river's surface but not much of its depth. A Frenchman discovered the spot, saw a painted pole planted on the bank and so named it. At the time of my visit, it was the thirteenth busiest port in the United States, a refinery town for the wells stretching south across the Louisiana wetlands and on into the Gulf of Mexico.

Exxon Wilmington was loading for her regular run to New York from tanks in Exxon's huge facility on the east bank. The plant looked and smelled like a city of the damned, all tubes and towers, flares and plumes. Hoses the size of sewer pipe linked the city to the ship, pouring aviation gas, alcohols, lube oils, and plasticizers into her tanks through about eight miles of pea-green piping running in ranks and files the length of her pea-green deck. Overall, *Wilmington* is 635 feet long, with a white sail of superstructure rising up five decks to the bridge and a great single stack aft. By the time we left, she would draw almost forty feet in the river, a bit less in the salt water of the Gulf of Mexico.

Wilmington's master, Larry Wonson, is young and carries his responsibility and authority pleasantly. "There isn't a product we carry that under the right conditions isn't explosive," he told me. His job is to make sure those conditions stay wrong — his explosives stored in mixtures too rich or too starved for oxygen to burn — and to get his ship through the Gulf, around the Florida Keys, and up the Atlantic Coast with as much expedition as he and she can muster. His chief mate, a Yugoslav with sad eyes and a handsome, accipitral face, handles the cargo. Wonson and his other mates — one looks a little owlish — busy themselves around the banks of electronic aids set in the bridge console and drive their beautiful, pea-green boat.

Late in the sizzling afternoon we dropped down the river. *Wilmington*'s big bowthruster kicked us away from the dock, and two Exxon tugs eased her around into the current. We ducked under the Baton Rouge bridge and headed into the first of endless curves. In the religion of the strong brown god, the shortest distance between two points is a meander. Barges lined the curving banks in the evening like boxcars in a marshaling yard. And at four the next morning, we passed New Orleans.

A ship's bridge can be a comforting place at night: quiet, the cricketing of instruments, darkness gone visceral in the dim red of the night lights. And outside, the shine of the city passing.

The pilot talked us through the sidewinder curves, his voice as low as if he were sitting in a theatre: "Right ten."

"Right ten," from the helmsman.

"Ease to five."

"Ease to five."

We dropped the Baton Rouge pilot at New Orleans and took on another for the hundred-mile run through the delta to the river's mouths. For a while we had both pilots on the bridge, and Larry Wonson and the two of them fell into that gentle joshing in which the game is to lay a fond hand on what the other fellow perceives as his weaknesses, and squeeze a little. With the Baton Rouge pilot, it was his age. About mine, late fifties, not too old for pilots in New York harbor but maybe getting on by the standards of this river.

I remembered Mark Twain writing of his apprenticeship here. "I haven't got brains enough to be a pilot," he complains. "And if I had, I wouldn't have strength to carry them around, unless I went on crutches." And Twain's instructor, his lord and master on the river, hollering, "Now drop that! When I say I'll learn a man the river, I mean it. And you can depend on it, I'll learn him or kill him."

The New Orleans pilot was a whiz. Three times he lay the ship in her marks. Three times the navigational signs on the banks lined up fore and aft, showing that we were heading precisely where we should. "Kind of surprises me," the pilot said. "Shows how much practice I've been getting."

"I'm impressed," Wonson said. "That must be what you're doing with that plastic boat in your bathtub."

By late morning we were well down the Southwest Pass. Oil well platforms stood in clumps off to the east like so many Godzillas at parade rest. Red dredges pumped silt from the bottom, and barges loaded with riprap lay along both shores. Cranes on other barges emptied the stone onto low levee walls. "That's Missouri rock," said the pilot. "Tell by the color." When the levees are finished, the water now leaking out across the spit of delta on both sides of the pass will be confined enough to keep the channel scoured. "Wait'll you see that in a good high-water year," said the pilot. "Some old tows won't make it against that current."

By noon we neared the end of the land. A large bulk carrier lay crushed against the eastern seawall, thrown there by Hurricane Juan. By a quirk, only one man had been lost. Two Cajun fishermen got on the radio. "I'm coming out after church" — a warmth of South and French in the voice — "and what kinda sea you got?" "Oh, not rough, not rough." The pilot called the churchgoer and ordered a bunch of shrimp from him. "He always gets the big ones, six to the pound," he told Wonson. "Like eatin' lobster." The pilot boat came in and kept station right under the port Jacob's ladder. Wonson gave the pilot a cheeseburger from the galley to tide him over until the shrimp. The pilot hustled down the ladder, and we turned into the Gulf of Mexico.

All the afternoon and well into the night, *Wilmington* bore away to the southeast, heading for the Straits of Florida, the pass between Florida and Cuba and then between Florida and the Bahama Banks. Her speed was twice what it had been in the river, and sometimes the great screw kicked up such a resonance that the ship shook as if she were laughing to herself. She, and we, were entering the domain of the Gulf Stream system, the blue god.

There, off the great submerged apron that Florida pushes out to the west and south of its thumb, we began to feel the god's might — the first pull of the Florida Current. It is a fire hose, a great jet that races in its core at speeds of about five knots. Through the Straits, around Key West, and up past Miami, past Jacksonville, Georgia, the Carolinas, to the bend of slim barrier islands that form Cape Hatteras. More than thirty million cubic meters of seawater flow in the Florida Current every second, a multiplication of Mississippis. The flow increases downstream: the water transported by the Gulf Stream off the Canadian Maritimes is on the order of two hundred million cubic meters — two hundred million bathtubfuls — per second. That far exceeds the combined flow of all the rivers of the world: the great-grandfather of waters!

The crew of the *Wilmington* called the skipper the Gulf Stream Guru. Larry Wonson minored in oceanography at Maine Maritime Academy and had been on the East Coast run for most of the twelve years he'd been working for Exxon Shipping. When the company started a program to use the Stream's current to cut fuel consumption and save time on the voyage north, Wonson worked with the marine scientists assigned to his vessel. He yearned for the satellite communications found on some navy warships and civilian research vessels. He did have radio reports from the Coast Guard on Stream location and velocity, and he had a Thermo-Fax machine drawing charts of the current for him. But he did most of his hunting by remembering the old lairs of the blue god and by using his senses.

"Look for a tide rip in about ninety minutes," Wonson told the mate, "setting southeast to northwest." That was one of his markers. Another was the west wall, the inshore edge of the Stream. Once he had it, he set a parallel course about a dozen miles out to sea. That way, he would probably be pretty close to the axis, where currents run the fastest. With summer coming, the west wall would likely be closer to shore and the axis currents faster. Maybe.

Even before we dropped our pilot, Wonson was figuring how to use an extension of the Gulf Stream that scientists call the

Loop Current. The Loop is like an aneurism, poking up through the Yucatán Channel between Cuba and Mexico and then bending around east and then south again, pulsing occasionally, shedding eddies that swirl off to the west and southwest. The current is strong enough to cause upwelling of deep rich water onto the shallower bottoms of the Gulf's best fishing grounds, and when that happens the result sometimes is a red tide, a bloom of tiny organisms called dinoflagellates. The dinoflagellates produce neurotoxins that shellfish can ingest in concentrations sufficient to cause paralytic poisoning in humans. When the organisms die in their billions, their decomposition uses up so much dissolved oxygen that bottom-dwelling fish cannot survive. So fisheries people want to know about the Loop Current, and government satellite people make a point of checking it out from time to time — not as often as they do the Stream proper, but well enough to give Larry Wonson a fair fix.

Wilmington had steamed along the eastern edge of the Loop the night before. "Many skippers fight the fast southern set," Wonson said of the current's flow. "They're always steering at an angle." Exxon captains run well south of where you would normally head east into the Florida Straits if you were interested simply in nautical mileage. Ten or fifteen miles below the shortest-distance course, they pull out of the Loop, which is by then running almost due south, and shape up for Tortugas. *Wilmington* did that and saved half an hour, maybe a little more.

When you're a hitchhiker at sea, it's best to stay on the bridge if the captain will permit it, out of the way, in some corner where you can watch and be forgotten. Wonson let me take over a high chair and small desk way over to starboard, and there I sat, eighty-six feet above the sea and almost five hundred behind the bow, across those miles of pea-green pipe. A perfect place for my purposes, perfect for jotting and reading and sending the mind to settle on the sea.

We ran past noon and on east, and as six bells sounded we picked up the light at Dry Tortugas, looming over the sea shimmer. From then on, and for miles down the Stream, we were locked in a special relationship with the land. Wonson wanted to stay in close enough so he didn't lose the extra safety factor of the navigational lights strung along the keys and up the Florida coast, but he kept the ship and her alcohols and her aviation gas well out from the wicked knuckles of the reefs. These shores eat ships. "If the Tortugas let you pass," the early sailors said, "beware the shoals of Hatteras."

Forty miles farther along the edge of the Florida apron lay the Marquesas, low and buggy, the habitat of a former poultryman who is probably the most successful treasure hunter in modern times. Mel Fisher has spent a generation looking for the wreck of the *Nuestra Señora de Atocha*, a Spanish galleon caught by a hurricane in 1622 and dragged to her death over reefs to a point south and a bit west of the Marquesas, her cargo of treasure scattered miles to the northwest. A decade went by between Fisher's first discoveries and the big find, a wall of silver ingots lying under sand. Tens of millions of dollars in silver, gold, and precious stones have been brought up. For this, Fisher has paid years, millions of dollars in bills and legal fees, and the lives of a son and daughter-in-law drowned when a workboat capsized one night at the dive site.

The *Atocha* — and the thousands of other galleons, merchantmen, privateers, and smugglers' and pirates' vessels wrecked along the keys — had been doing pretty much what *Wilmington* was doing: catching a ride on the Stream. The Spanish treasure fleets came out of Havana with bars and ingots of silver as their principal cargo, with gold bars and ornaments from Mexico, pearls from the islands off Venezuela, emeralds from Colombia, much of it listed on the manifests but much of it stowed as contraband to fatten the wallets of proud and thieving *hidalgos*. They ran a route called the Carrera de Indias, and it was as solid as an interstate: north to the Florida shelf; up the chain of the keys, keeping the land in sight for the benefit of their navigators and the digestion of the ships' companies; past the threats of the

Bahama Banks, then the run toward Bermuda, the Azores, and at last Spain. Returning, the fleets sailed south until they could pick up the northeast trade winds and followed them, and the Equatorial Current, west to the Caribbean.

Wilmington stayed just out of sight of the keys as the sun dropped. By four on Tuesday morning we were off Fowey Light, south of Miami. By midday we were off Jupiter Point and making 20.5 knots, as against 16.5 in slack water. Larry Wonson grinned and said, "I sense the presence of external assistance." Nothing showed in the water to indicate current except some windrows of weed from time to time. We could have been on a protected inland sea.

History loves a choke point, a narrowing, like the Straits of Florida. Christopher Columbus might have lost his dream, his ships, perhaps even his life here. Samuel Eliot Morison, the great historian of the European discoveries of the American continents, wrote that if the Admiral of the Ocean Sea had not decided to alter course and follow flocks of migrating birds as he neared his first landfall, he and his mutinous crew might have sailed into the grip of the great current. Then, Morison speculated, "the fleet would have touched (and perhaps more than touched — gone ashore) on the coast of Florida, somewhere between Jupiter Inlet and Cape Canaveral." If they survived that landfall, they might well have been "swept along the coast of Georgia and the Carolinas, returning to Spain (if they managed to return) by the westerlies north of Hatteras and Bermuda."

Juan Ponce de León, who sailed with Columbus on his second voyage, did land near here. He was looking, among other things, for certain restorative waters rumored to be found nearby. What he found — the long beach and the low scrub behind — he named for the season, Pascua Florida, Easter. And what he found in the process was the Gulf Stream, which whisked one of his vessels out of sight before it could escape and return to anchorage. Ponce de León turned south, following the inshore countercurrent, past the keys, which he called the Martyrs, "because the high rocks at a distance look like men who are suffering."

It was not Ponce de León but his pilot, Antón de Alaminos, who first put the Stream to use for European civilization. When the conquistadors took Mexico, Alaminos remembered the blue god. His mission was to bring word of the conquest to Spain. He had the imagination and fortitude to bear away down the great flow and in so doing established the Carrera de Indias. Along the way, in all probability, he made a rutter, a set of written sailing directions.

Early navigators took proprietary pride in making rutters and keeping them from the competition. Richard Hakluyt, the sixteenth-century chronicler — and booster — of British maritime voyaging, collected rutters, many carrying the poetry of things closely observed, many from the Spanish:

> If from Havana thou wouldst set thy course for Spaine, thou must goe Northeast and shalt so have sight of the Martyrs, which stand in 24 degrees and a half [of latitude]. And the coast lieth East and West. The marks be these, it sheweth like heads of trees, and in some places certaine rocks with white sandy beaches.

The use of the lead was all-important in the shallows, not only for depth but for reading the bottom. Black sand sticking to the tallow packed into the end of the lead meant one location, "shellie ground and periwinkles" another.

Larry Wonson had a rutter of his own. "Even though the technology is there," he told me, "if you're going to pursue the Gulf Stream as a skipper, it's got to be mostly through your own experience." Once you get up even with the northern end of the Florida peninsula, he said, it gets harder to stay with the Stream. So you look for weed lines, for squalls building up along the western edge. You read the color of the sea, trying to stay where indigo predominates. You check the temperature of the seawater in the engine intake. Most of all, you check your speed, looking for that external assistance.

We were north of 30 degrees latitude now, moving out of the first of the three great bands of wind and weather in the Northern Hemisphere (mirrored in the Southern). We were leaving an ellipse of air called the Hadley cell, which rises in the doldrums just north of the equator, moves poleward at high altitudes, sinks into the horse latitudes, and comes back toward the equator. Winds were light and fickle seaward of our present position, but in a while, as we steamed up the coast, they would become westerlies, the winds Alaminos was banking on, marking a second cell of atmospheric circulation. Well north, above 60 degrees, the winds turn easterly again in the third cell.

Our weather was gentle and fine, but during a recent trip north Wonson had run into a high countering wind out of the northeast and a huge swell from an offshore storm. *Wilmington* moved inshore, where Wonson was sure the riding would be easier. The Stream did knock perhaps fifteen feet off the seas, but they were still running at twenty-five feet, and their angle of attack forced him to return to the current. Wind and water did brutal battle there. *Wilmington*, a tender ship, rolled like a log. Nothing of consequence got bent, but the crew took a beating in the sharp water, and *Wilmington* lost a lot of time.

The hitchhiker ate varied and wonderful food. He sat on his perch in a sea reverie. Three cadets from maritime academies in Maine and Texas, along for the training, practiced shooting the sun and other navigational procedures out in the heat on the starboard bridge wing. One of the two third mates aboard commented on the decline of the American shipping business. "The romanticism is out of it," he said.

"Do you realize you're driving one of the largest machines man has ever made?" asked a cadet.

"Thrilling," said the mate.

Time disconnected. Swells slipped across the bows in close order, mere ruffles from this height, following on in a rhythm slow as sleep. *Wilmington*'s three thick, stubby masts, crowned with batteries of working lights, became fighting castles on an ancient ship, and I entered waking dreams of great galleons pushing north, of Celts, a thousand years before Columbus, caught by the current, dying of thirst and scurvy, the few surviving until chance put them ashore near the mouth of some sweet New England river. There are evidences — arrangements of rocks that some say just might be Stonehenges — near some of those rivers. Hakluyt believed in pre-Columbian voyages. He wrote of Madoc, the son of Owen Guined, prince of North Wales, who "sought adventures by Sea, sailing West," until he came to a land which "must needs be some part of that Country of which the Spanyards afferme themselves to be the first finders. . . . Of the voyage and returne of this Madoc there be many fables fained," Hakluyt said, "as the common people do use in distance of place and length of time rather to augment than to diminish: but sure it is there he was." And then the punch line: "Whereupon it is manifest that that country was by Britaines discovered long before Columbus led any Spanyards thither."

Perhaps. Kenneth Emery and Eleazar Uchupi, scientists at Woods Hole, took time in the introduction to their study of Atlantic geology to discuss the wanderings so many historians dismiss as fables fained. They refer to markings on stones, cliffs, and cave walls that have been deciphered sufficiently to indicate the presence of Celts, Iberians, and Libyans in the Americas hundreds and thousands of years ago. "When these various findings become better organized and supported by other workers," they say, "they may completely revise the concepts of ancient travel and early visits to the New World." So dear the possible.

How many hundreds died in the Stream during the discoveries and rediscoveries of the American coast? How many, for every one who stepped or crawled ashore? How many vessels, dismasted, drifting north and east with the current, the castaways lolling? We are not comfortable with such suffering. Winslow Homer learned that. When his painting *The Gulf Stream* went on exhibit in the early years of this century, its depiction of a black man lying helpless on the deck of a derelict sloop caused enough exclamations of public distaste to bring reaction from the artist. "The criticisms of 'The Gulf Stream' by old women and others are noted," Homer wrote a dealer who had told him about the flap. "You may inform these people that the Negro did not starve to death. He was not eaten by the sharks. The waterspout did not hit him. And he was rescued by a passing ship which is not shown in the picture."

Larry Wonson broke my trance: "The west wall!" He had his glasses on something to port, a slick on the surface running parallel to our course, perhaps two miles off. "If we were to cross the edge head-on," Wonson said, "you'd see your wake in the distance moving north with the Stream and your near wake moving south in the countercurrent." We were inshore of the zone of maximum velocity, but radar showed a lot of traffic there. Wonson didn't want company and the headaches of constant course changes, so he stayed west of the fast lane.

By nightfall the Stream's influence was weakening. From a solid four knots at Jupiter, the current under us had slowed to two knots. Still worth the ride, but the blue god wasn't in champion form this trip. We wouldn't set any records, nothing like the three days and seventeen hours logged on a New York run last year.

I have never seen such quarters at sea. My stateroom had its own head, a double bed, a huge square port, an easy chair bolted to the deck, a desk, and a large alarm box on the starboard bulkhead. Every night so far, the lighted squares on the box had carried one message: Engine Room Unattended. *Wilmington* is so sophisticated that there is no need for humans to lose sleep tending her 17,000-horsepower diesel. But in the dead of one night, the next to the last of the trip, the box began to scream. Blasted from deep sleep, I responded in kind. A different message glared at me: Fire and Smoke in the Engine Room.

To the bridge! Perhaps the demon bells weren't ringing there. They were not. Pure serenity in the womb-red of the night lights. The mate whispered into the telephone. The engineers, he said, couldn't find anything wrong in the engine room. A second call. They had located a small leak in the exhaust system. Wonson arrived, in bare feet, and brought the ship up short. The engineers protested his sudden dumping of the load off the engine. Things like that can cut engine life. "I'd rather run that risk than have an explosion," Wonson said.

In fifteen minutes the engineers called to say they'd repaired the leak. I passed one of them, tall and confident, on my way back to my king-size bunk. "Just a small leak, all fixed," he said, grinning down at me. I thought of the aviation gas and the alcohols all nestled in their tanks up forward and bobbed my head like a robin.

With the day came heavy rain squalls, flattening the small seas. We were closing now on Cape Hatteras. All along Florida and the Carolinas, the Stream lies against the continental slope, where the bottom drops down from the gentle declivities of the continental shelf. It generally follows the slow curves of the one-hundred-fathom line and runs surface to bottom. From the Straits of Florida to Hatteras, the Stream flows across the Blake Plateau, an ancient coral reef lying eight hundred meters, more or less, below the bright surface. All across the Blake Plateau are signs of current — scourings here, valleys there. Thanks to the Gulf Stream, the plateau is about the only place in the North Atlantic where you can find fairly large deposits of manganese nodules, those mysterious concentrations of valuable metals that litter the floor of the deep Pacific. The blue god sculpts and sweeps. He bore away the wreckage of the space shuttle *Challenger*, making a difficult recovery operation almost impossible.

At Hatteras, the Gulf Stream literally falls into the North Atlantic. In ways that scientists still don't know much about, it rolls off the Blake Plateau into the smallest, warmest, and saltiest of the world's seven oceans (North and South Atlantic, North and South Pacific, Indian, Southern, and Arctic). Here is the Inlet, the nozzle of the fire hose. Downstream (somehow I find it difficult to remember that down the Stream is up the coast) the current broadens and begins to corkscrew into meanders that grow into what on land would be called oxbow lakes but here form rings or eddies inshore and seaward of the main flow. More importantly — and more mysteriously — the flow itself heads offshore, a little north of east, looping along toward the tail of the Grand Banks. It is the true Gulf Stream now, not the Florida Current. It begins to transport more water, and its reactions to gravitational and rotational forces become more complex.

I never made the trip up the long foredeck to the bow. I love standing over the water, looking down as the blade of the hull slices and plows. Often dolphin surf there, planing off the bow, heads slightly down to catch the angle of sheared water. But something — the piping, the cargo — put me off, and I stayed astern and thought of dolphin. An animal well studied, just as the North Atlantic is the most thoroughly studied of oceans and the Gulf Stream the most thoroughly studied of its currents. And yet the more we know, the more we know we don't know. Mystery still outswims certainty in the waters under *Wilmington*. So, said I, if I were to spy a sea serpent right now, I would be within my rights.

Something like a monster was seen in December of 1947. The Grace Line's *Santa Clara* was steaming off the Carolina coast on her way to the Caribbean port of Cartagena when suddenly her watch officer saw something huge to starboard, about forty-five feet long, with an eel-like head and a body about three feet through. The thing kept coming and the ship ran it down, leaving it thrashing and pouring blood in the wake. Earlier, in 1913, a steamer reported making contact with a sea giraffe. It had a twenty-foot neck, blue eyes that "took in the ship with a surprised, injured and fearful stare," and a wail like a baby's. Sightings like these can unsettle the respectable. Thoreau swears that Daniel Webster once saw a huge sea creature near Plymouth and cried to his companion, "For God's sake, never say a word about this to anyone." How far we have wandered from the Book of Job, from Leviathan: "Who can open the doors of his face? His teeth are terrible round about . . . his eyes are like the eyelids of the morning."

A lot of sightings have turned out to be huge basking sharks, alive or dead. Baskers are fairly common at these latitudes, though the Stream itself, with its relatively small populations of plankton, shouldn't be a favorite hunting ground. The true hitchhikers on the current turn out to be pelagic racers like the bluefin tuna. Or the involuntary riders, the plankton: winged snails, strings of glasslike barrels called salps, the flat ribbons of Venus's girdles, comb jellies. Most are weak swimmers, there because the Stream caught them up.

On the last morning we sighted the coast again, the highlands of New Jersey. Above them and all along the western horizon a brown torrent flowed, a Mississippi of the air, a river of smog. Larry Wonson told me we had left the Stream about eight hours before. Sometimes, if you're lucky, a meander of the current will give you a ride inshore; in winter that can mean not only a boost in speed but a long, looping buffer against icing up.

No such boost this time. "The engines went 1,595 miles," Wonson said, "and the ship went 70 miles more than that," in the push of the Stream. The current had saved us about four and a half hours, only a little more than half the cut in time for a good trip. Still, in fuel and related costs, a cut of four and a half hours in steaming time worked out to a saving of five thousand dollars. That, multiplied by the run of two ships — *Wilmington* and her sister, the *Exxon Charleston* — every two weeks or so adds up to real money, even in Exxon eyes.

We were well within the inbound traffic pattern now, ambling in a line of wide-hipped container ships, tankers, and bulk carriers. Ambrose light showed ahead, and the pilot came aboard bringing the traditional pilot's gift, the local papers. I stole a riffle through the *Times* and there, looking as if he were going to cry with delight, was treasure hunter Mel Fisher. He held a glass bottle, maybe an old pickle jar. In it were more than two thousand emeralds brought up by his divers from the wreck of the *Atocha*. A spokesman for his outfit, Treasure Salvors, was quoted in the caption: "It's the volume that's incredible. You're talking about multi, multimillions of dollars worth." Treasure types spend so much time not finding anything that they tend to gabble when they do. But clearly the volume was incredible.

Wonson told the pilot he wanted to find a spot in the Stapleton Anchorage, in the lower bay near Brooklyn, that was clear of everybody else. The pilot got on the radio as we came in under the belly of the Verrazano-Narrows Bridge. We backed and filled, using the bowthruster to get us right with the current, and then Wonson dropped his hook. In a short while a tug would bring a barge alongside, and *Wilmington* would pump it full of this and that, cutting her draft by several feet so she could maneuver in the tricky curves of the Kills on her way to make her deliveries back of Staten Island.

A water taxi came to take a few of us off. One was a young fellow with a guitar, resplendent in the hot air in his white jeans and T-shirt. He had finished his tour of sixty days and was going ashore for forty-two. Up to Maine, he said, grinning a little like Mel Fisher in the AP photo. Larry Wonson would stay aboard. No chance to duck up to his native Rockport, north of Boston, and lift a few with his old fishing buddies. He'd spend a couple of days unloading and then head south.

The blue god would not befriend him on the return voyage.

Wonson had a choice. He could ride the countercurrent down inshore of the Stream, or he could go outside. The trouble with the inside route was that Wonson couldn't get as close as he'd like to Diamond Shoals to ride the knot-and-a-half favorable current there. Exxon doesn't want to risk grounding and keeps its skippers twenty miles off places like Diamond. So *Wilmington* would go outside.

"I like to cross the Stream below Diamond," Wonson said, "at ninety degrees, as painless as possible. Then you can get on those cold-eddy currents, counterclockwise, for maybe three or four hours." Those are the rings of continental slope water caught in the Stream's offshore meanders and pinched off to whirl slowly and powerfully, south and then often west, to rejoin the mother current. "But," Larry added, "you can't predict where you're going to find those eddies." Or what part of the whirl you'll hit. One time, on her way up the coast, *Wilmington* turned right, out of the Stream, to dodge a hurricane and ran smack into an eddy current. "The ship was clicking along at sixteen and a half," Larry said. "Then, bingo! She's making thirteen."

The taxi cast off and began its five-minute scuttle to the Staten Island ferry terminal. I took one last look at the scuffed black hull, the pea-green piping. The skipper and his chief mate stood out on the bridge wing talking, probably about the complexities of unloading which product from what tank. Far up the harbor, the stelae of lower Manhattan rose through the airborne Mississippi to clear light, and I thought of *Wilmington,* outbound, hunting countercurrents that would give her a ride back to the strong brown god, to a proper river.

3
The Admiral Imagined

ON THE NORTH ATLANTIC subtropical gyre, as on other great oceanic circulations, man is as much a migrant as are tuna and eel. He came to it later than they, and has his problems with it still. Though he puts out to sea, he does not belong there and, unless he goes back to gills, never rightly will. Still, he migrates.

It is wonderful fun to speculate on the first attempts, to tell the story of Saint Brendan or Madoc or the Phoenicians and then smile and shrug. Hard evidence is still in small supply. Reference over centuries to St. Brendan's Island or Antilia, the island of the seven lost cities erected by Christian bishops "opposite" Portugal, is but the magic of folklore. Roman coins have been dug up in American fields, but that doesn't necessarily mean some wandering caesar came this way. European, African, and Oriental faces gaze out from among the Amerinds in the collections of small clay heads gathered from pre-Columbian graves in Mexico. They entice but do not, by themselves, prove. Only hearths and carbon-dated middens, like those at the Norse settlement high on the northern finger of Newfoundland island, settle the matter for science. Meantime, we can take the pleasures of speculation.

We can speculate, for instance, about the Portuguese. We know that through most of the fifteenth century they were the masters of the North Atlantic. Their captains were catching up with the great maritime traditions of the Arabs and the Chinese (whose Ming rulers, recoiling from contact with foreign cultures, were about to turn their backs on the sea. "If political changes and cultural endogeny had not stifled the ambitions of Chinese sailors," writes ecological historian Alfred Crosby, "it is likely that history's greatest imperialists would have been Far Easterners, not Europeans"). Portuguese ships coasted Africa, doubled its great southern cape, and went on to India and the Orient. They sailed far north and went west to the Newfoundland fisheries as the century closed. Historians place them in Brazil by 1500. Given that proficiency, isn't it possible that Portuguese visited the Americas well before that? And once that question is out, the way is open to: Isn't it probable?

I flew to Lisbon on Halloween of 1986, not to celebrate any primacy of the Portuguese but to sail from a port in the southeastern corner of the country on the topsail schooner *Welcome*, Arthur Snyder owner and master. *Welcome* was bound for Madeira, La Palma in the Canary Islands, and along the southern arc of the gyre more than 2,500 nautical miles to Antigua, at the gates of the Caribbean. I knew something of Portugal's *marinheiros* from some of the books I carried with me: Samuel Eliot Morison's histories of the early Atlantic crossings and J. H. Parry's *The Discovery of the Sea* among them, and an English translation of excerpts from the journals of Christopher Columbus (who greatly respected Lusitania's skills on blue water). In Lisbon and on the way south down the profile of the Iberian peninsula, I could sense how that seamanship might have taken hold.

In the last medieval years, Portugal was a country pauperized by inflation and the perennial fighting with Castile, and wasted by the Black Death. The land was not rich enough to feed even a decimated population and provide a surplus for trade. The ocean was rich but too powerful along the open Atlantic coast for fishermen to fight more than a few leagues off the land. Against that adversity, the only way lay farther to the south. There might be gold from Africa, and perhaps an alliance with the Christian king Prester John. A new route to the spiceries of the east. Good land for sugar and other commodities to set northern Europe adrool, if such was to be had among the mythic archipelagoes of the Atlantic.

In following this imperative, Portugal had to borrow manpower and money, mostly from the Genoese, who had a thriving colony in Lisbon, because there simply weren't enough native seamen or caulkers or sailmakers or shipwrights around. But the venture advanced, the kings of the country nurtured it, and men came forward to make it their own, men like the half-English Henrique, son of João I, known as Henry the Navigator.

Henry still lives among the hills of Lisbon. His name is everywhere the symbol of the time when Portugal was the greatest empire in Europe, with colonies in Africa, India, South America, and the Far East. I saw him on the shores of the Tagus estuary, called for its color the Sea of Straw. The sun was beginning that fast slide that shows us the spin of the earth, and the light washed his statue and went gray. Henry is the figurehead on a great cement prow, and behind him are the captains, caulkers, and capitalists who followed him in life. In Henry's hands is the model of a caravel, the small and sweetly seaworthy vessel that gave Portugal her way out to power.

In her Mediterranean form, before her marriage to the square-rigged, clinker-built ships of England and northern Europe, the caravel was rigged fore and aft, with great triangular lateen sails that came down to her from the Arabs. She was a great coaster and could sail fairly close to the wind. With the addition of square sails, she could also run before the wind. And that, in Iberia's westering, made all the difference. Spain built caravels and so did Lisbon, many at a spot less than a mile up the estuary from where Henry stands contemplating the tan water.

The road to Vilamoura and the schooner *Welcome* heads south out of Lisbon, crosses the Tagus, unreels on plains, and runs in tangles through dry ranges. It roves through the farms of the Algarve, past plowed land and cork trees bared of their bark from ground to first forking.

Towns sprout stucco and red, their fortress castles heavy on the hills. Moors lived in this country for centuries. At places like Alcácer do Sal there were also Greeks. In those alleys behind the pale walls lived men who knew mathematics and astronomy and therefore had knowledge that could be used to direct a ship upon the open sea. The Christians, interested far more in souls than in star angles, had remained only vaguely aware of that knowledge since the burning of the great libraries at Alexandria. When they advanced against the Moors, they found themselves too ignorant to understand the culture they were conquering. A few Christian "scholars," working with Moorish tutors, painfully translated the pagan texts into Latin. And in Iberia, dawn came to the Dark Ages.

After a few hours' drive, the road runs out to the end of the world. Cape St. Vincent, its high bows of limestone plunging southwest, was the western limit of the known land for the cartographers the early discoverers studied. Beyond lay behemoths and one-eyed cannibals and, possibly, India. It is said that Henry the Navigator came here from his holdings in Lagos, the port a few miles east in the lee of the cape (where the young Christopher Columbus swam ashore, wounded in a sea battle between his fellow Genoese and a Franco-Portuguese fleet, to begin a long residence in the country). Tucked under the endless wind in a cup of rock, smelling rosemary and hearing the sea's voice in its caves under the cliffs, I thought of the prince. He was a knight of God. He was a knight of commerce. He wished to find Prester John, to save souls, to turn a profit. A man of his time — a time when the old certainties were falling away and followers of Christ were turning their contemplations from heaven to earth.

From my lee I could look across an embayment to Sagres, the town where Henry is supposed to have put together his version of an oceanographic institution. It is not at all clear that he did so, though the historian Parry says he had a fortified village built there "for the protection and convenience of ships waiting off the point." Nothing remains but a village and the dun tableland and the sea.

East I went, beyond Lagos, to the country of the new merchant princes, a land of giant water slides for the delectation of the foreign customers; condominiums with the pitch on the huge sign in front, in English, "The Big One Gives You More"; boutiques and restaurants and a golf course. Then Vilamoura, a self-sufficient resort walled away from the Algarve by design and money. Many British come here, running from the gray of their winters.

The marina, a manmade bowl, lay in dusk, hedged with masts. *Welcome* stood out like a fine saltbox in a row of trendy split-levels. When I lost sight of her big white hull, her yards guided me along the floats to her slip and to her master, Arthur Fenimore French Snyder. *Welcome*, out of New Bedford, Massachusetts, had come all the way across the North Atlantic with the Gulf Stream extension to Scotland and then here.

Well before the American Revolution, some anonymous soul in the fishing port of Gloucester, Massachusetts, is supposed to have remarked on the progress of a two-masted, fore-and-aft rigged vessel at play with a fair wind. "See," he is supposed to have cried, "how she scoons." Depending on the source, *scoon* meant to scud, to fly downwind, or simply to glide. Schooners can scud, particularly if they are fitted, as *Welcome* is, with square sails on the foremast. But a schooner wind is one coming in aft of the beam, putting the boat on a broad reach. Running or beating, schooners are among the most adaptable vessels under sail. Fishermen on the offshore banks loved them, and so did coasting captains. Americans built them every which way, from baldheaded rigs with no topmasts to great six-masted lumber boats.

Welcome says a lot about Arthur Snyder. The name comes from the vessel that carried William Penn to his grants in Pennsylvania. Snyder is of Penn's faith and of his city, a birthright Quaker from Philadelphia. He is an embodiment of the Quaker joke about going through life doing good and ending up doing well; at sixty-eight he is vice chairman of a solid Boston bank. Unlike many Friends, who forswear taking up "external arms" against their fellows, he fought in World War II as a navy deck and engi-

neering officer, going in before Pearl Harbor and coming out well after V-E Day.

Snyder's schooner is at once a decade and a century and a half old. She was built, in the famous Concordia yards in Massachusetts in 1975, from plans adapted from some drawn up in 1815 for a revenue cutter, the ancestor of the Coast Guard cutters that hunt today for drug smugglers and weekend sailors in trouble. In concept she is of an age with my great-great-grandfather, Moses Hillard, a farmer-seaman. He captained the 232-ton *Amiable Matilda* and other vessels sailing out of Mystic, Connecticut, and New York. A French privateer captured him in the West Indies when he was nineteen, and when he was one of the most respected masters on the North Atlantic, the French diddled him again. Friends of Napoleon Bonaparte approached Moses after Waterloo and asked him to carry the defeated emperor to safety. Hillard had a false bottom built under a water butt and lay off the appointed headland at the appointed hour, but ashore all was silent.

Welcome is built of wood: cedar on oak in the hull, and eastern pine sheathing agleam belowdecks. She has none of the banks of mechanized winches that marked the upscale wonders around her. The heavy work is done by hand, with block and tackle. It takes a professional to master her, and Snyder is that. He figures he began sailing in the womb. This trip, he had five souls to help him: his daughter Carrie, slender and competent; a west-country Briton named Mike Jones, an Atlantic sailor who could cook in several cuisines and handle a watch, all with detached aplomb; and three amenable items of movable ballast: a two-person film crew and me.

Welcome is about forty-five feet on deck and thirty-six at the water line — beamy but well proportioned. Snyder was a bit like that himself, somewhat roomy amidships when I first saw him, but he told me he'd lose his business-induced protuberance to the isometrics of the sea, and he did. His face is just right for his humor, which is blustery, often to a point approaching, but not quite reaching, insensitivity. He can threaten, deadpan, and then mug and let go with a satisfied and satisfying laugh. His anger is so quick it boils away before it does much damage.

The five-hundred-mile run to Madeira began on a warm midmorning in early November. Snyder took his departure from the light at the entrance to Vilamoura's marina as carefully as thousands of captains over the centuries have taken theirs from the Navigator's headland. The procedure may be less important in our electronic day, but it is still observed. The point you depart from is the anchor from which you unreel the line of your voyage. The early *marinheiro* might not know where he was, but he had a pretty good idea where his departure point lay as he measured latitude, and so stood a fair chance of raising it as his last and best landfall.

Snyder took us out from the lee on *Welcome*'s diesel. Far off, a dun shadow to starboard, St. Vincent loomed and slid under. When the northerly breeze had freshened to his satisfaction we raised the plain sails, the everyday fore-and-aft equipment, and *Welcome* began a discourse with herself that was to last the full crossing. Masts coughed and creaked in their steps, wooden members muttered, and the sea, cut wide by the bow, ran astern with a dry hiss.

I could see then where *Welcome* comes from. Her mainsail and foresail are truncated lateen sails. Her forecourse and topsail are as square as any canvas found in northern Europe at the time of John of Gaunt. She is the daughter of the marriage between caravels and cogs. Her ancestors opened the Atlantic.

We made Madeira in a bit under four days, time Snyder used to set watches and educate the lubbers. We learned what to do if one of us were to fall overboard. (The luckless wretch should resist the tendency, strengthened by shock, to tumble in mum. The helmsman, alerted by the screaming if not the sight of his mate in midflip, must steer a true course so that the ship can return for her struggling son on the reciprocal. The watch should get the float and buoy and strobe light overboard this side of immediately.) We learned the location and purpose of a few of the more than one hundred belaying points aboard, where stays and shrouds and halyards and jiggers and sheets are secured. We learned how to brace yards, and scandalize the mainsail, and raise the forecourse. For days I studied an etching of a full-rigged ship in the after head and had some terms memorized — bonkwratches, poot-sledges, slugnabbles, jug nuzzles — before the suspicion of spoof dawned. Still, there didn't seem to be much distance between the imaginary bobsprout and *Welcome*'s very real cat stopper. Not to movable ballast, there wasn't.

We also learned how Snyder ran his boat. He worked himself fiercely, always roving, peering at rigging, getting out his ditty bag to repair a block or make a fitting, pumping the bilge and counting the strokes, shooting the sun and working through the numbers of his navigation. By dusk he showed the wear. But in the midwatch, from midnight to four, he came alive again. We would talk, whispering in the wind to keep from rousing sleepers by the companionway, about the state of the world. "I don't understand the Italians," Snyder would say. He would quote Byron in the silver air: "She walks in beauty, like the night/ Of cloudless climes and starry skies." Or Tennyson, his favorite. Or himself. Snyder's Eleventh Commandment: "Thou shalt not think that the Lord thy God is better than anybody else's."

For years Snyder has been at work on a compendium of nautical terms and beliefs, and he was glad to enlighten me. I had always thought the saying "Cold enough to freeze the balls off a brass monkey" was naughty, not nautical. No, he said. A monkey, on warships of old, was a brass rack for cannonballs; in severe cold the brass shrank more than the iron shot, and out they popped. In my innocence, I talked about sailing to Madeira. "Never *to*," he corrected. "Always *toward*." We lived by the sufferance of the sea, he said, and it was both arrogant and offensive to come right out and say where we were going. *Welcome*'s bow struck fire from the swell, flakes of green and living light. We were a streak of midnight bioluminescence 350 miles off the deserts of Africa. Yes, *toward,* not *to*. I watched the fire in the wake and remembered the men in my family who had been here before me. Benjamin Franklin Hillard, brother to Moses, was lost off the coast of Spain in 1829 at the age of nineteen. And another brother, George, died in 1830, on the island now thrusting its blackness against the night dead ahead. He was thirty-one.

Madeira is an unexpected island, all runneled spines and green spires rising to eight thousand feet with no running start, roofs of red tile on the fallaways and the sinister fortress over the harbor at Funchal. The Portuguese found Madeira as they found the Azores, as an afterthought. Exploring Africa was what they really had in mind. Sailing down that coast was fairly easy for the first *marinheiros*. They had the winds generally on their sterns and the extra half-knot push of the Canary Current, the easternmost flow of the North Atlantic subtropical gyre, all of which made it next to impossible for them to take the reciprocal course home. Instead, the more daring — and consequently often the longer-lived — among them tried heading northwesterly on a long reach until they came into the zone of the westerlies. There, they bore off to the east and good old Cape St. Vincent, or Cabo Raso near Lisbon. *Volta do mar*, they called it, the return by sea rather than by coast. In 1420 one returnee from the sea visited what is now Porto Santo, Madeira's twin. Seven years later, another raised the Azores, one third of the way across the Atlantic from Gibraltar to the Grand Banks. Whether these were discoveries or rediscoveries is eminently arguable, given the traffic in the area from Phoenician and Roman times onward.

The specter of Columbus boarded my mind at Madeira. He had married a Madeiran, my traveling library told me, the daughter of a captain of Porto Santo. Columbus studied the captain's charts and mementos. He listened to his brother-in-law talk of strange objects found offshore and on the beaches after storms, intricately carved wood and huge canes. "He was careful," wrote Columbus's son Fernando, "to treasure up whatever information relating to this matter he could collect from travellers or seamen. In this manner he came to a firm persuasion that to the West . . . there were islands which might be reached by sailing in that direction." Thus did the gyre tempt the Admiral of the Ocean Sea.

The film crew disembarked at Funchal, replaced for the crossing by two experienced sailors. Howard Freeman is one of Art Snyder's closest friends, a man almost hidden in gentleness. An inventor with more than twenty patents, Freeman developed the nozzles that produced mists heavy enough to drown fires aboard ships during World War II. Some of them were used aboard a troopship afire in the Atlantic. She was the *Wakefield*, and her engineering officer was Arthur F. F. Snyder. Neither knew the other until well after the war, when banker Snyder helped set up inventor Freeman in his own business.

William Claxton is a ruddy, blue-eyed, joyful trumpet blast of a man from Cornwall. He has sailed and fished those English seas and now serves as boatman for the Royal Cornwall Yacht Club in Falmouth. Willy, he's called, and his Norfolk accent at first mystifies the Yankee ear with its *ayes* for *ays*, its dominant *ohs*, and its disdain for *ares*. *Bone in Nofuk, mite. Propah plice.* Willy sailed this southern route to Antigua last year aboard a racehorse yacht. He has that knowledge, along with the tales and terrors of forty years spent on and at the edge of the sea. He cannot swim.

Welcome, replenished, sailed for La Palma, westernmost of the larger Canary Islands, some 280 miles to the south. Pliny wrote of this high and dry archipelago tucked close to the western Sahara, and it was rediscovered by the Portuguese well before they found Madeira and the Azores. Unlike the Madeiras and the Azores, the Canaries were populated. Columbus called the inhabitants "neither black nor white." They were Guanches, proud people, excellent mountaineers. They gave the Portuguese hell and the Spanish after them, signaling among the peaks with sharp whistles and hurling stones with appalling accuracy. It took decades for Castile to subdue them.

On his virgin voyage, Columbus sailed straight for the Canaries from the Spanish port of Palos, not far east of Vilamoura. It was an eminently sensible thing to do. Morison says the Admiral's charts (or what passed for them) told him that his initial target — Cipangu, or Japan — lay at roughly the same latitude as the Canaries. The practice of the times was to travel the parallels, though even the best navigators had trouble maintaining latitude. Columbus knew from the experience of the Portuguese that to sail due west out of Palos would be to buck the westerlies. He knew from his own experience that the northerly winds of summer's end would carry him easily down the African coast.

Once free of the Canary calms, he was pretty sure, he'd have a fair wind for Cipangu, though he could not have known that the trade winds would carry him all the way across the Atlantic. So, reasoned the Admiral: eight days' sail to the Canaries and the services I can get there from the subjects of my Castilian sponsors. Thence, according to his conviction, about 2,400 nautical miles to the land of the rising sun, and then a bit farther to Cathay. He was close: *Welcome*'s crossing was of 2,600 miles. But ah! the destination. Though Columbus died cherishing the belief that he had found what he had sailed for, the shortest route to Japan from the Canaries runs 10,600 nautical miles — across two oceans and one isthmus. For the Admiral, as for many of his age, the globe was not seventy percent ocean but eighty-five percent land. Did not the Bible speak of the Divine Architect, saying "six parts thou hast dried up"? Nature, said a Portuguese philosopher, "could not have made so disorderly a composition of the globe as to give the element of water preponderance over the land, destined for life and the creation of souls."

The presence, now so much in my thinking as our course merged with that taken a bit shy of five centuries before, was that of the magnificent failure, one of humanity's most poignant and powerful archetypes. Columbus did not know where he was going, but he was so lucky, so enduring, so sensitive to the sea that he got there. And once he had found land, no matter what he called it, no matter how many embassies he sent inland bearing letters from the sovereigns of Spain to the Great Khan ("We have learned with joy of your esteem and high regard for us and our nation and of your great eagerness to receive information concerning our successes"), he could find it again. He had a wind rose in his head, they said of him. That he also had a zealot's selectivity, relying on faith and conviction to shape a geography that many scholars of the day laughed at, for right reasons or wrong, was not in the end important.

The Admiral's son and biographer found the right epitaph in the works of Seneca, who himself was of the opinion that with a fair wind one could sail from western Spain to India in a matter of a few days. In Seneca's play *Medea,* he marked this passage: *"Venient annis/ Secula seris, quibus Oceanus/ Vincula rerum laxet . . ."* ("An age will come after many years when Oceanus will loose the chain of things, and a huge land lie revealed; when Tiphys will disclose new worlds and Thule no more be the ultimate"). Fernando, writing also in Latin, noted in the margin, "This prophecy was fulfilled by my father . . . the Admiral in the year 1492." Indeed. Tiphys was pilot to Jason and the Argonauts. Columbus was pilot to all Europe.

At no time in the crossing would we be so close to Columbus as here among these great rocks. There was Tenerife, way to the west, two and a half miles high, rising like a breast from the sea. Columbus wrote in his journal that he had seen a "great eruption of flames" from its peak. Hidden behind it lay Grand Canary, where the *Pinta,* which had been performing poorly in the following wind, had some of her lateen rig replaced with square sails. Nearby was Gomera, where the Admiral topped off his provisions and learned from "many respectable Spaniards . . . that they every year saw land to the west of the Canaries." St. Brendan's island, no doubt of it, San Borodon. "Probably," writes Morison in his *Admiral of the Ocean Sea,* "you could find aged fishermen in . . . Gomera today who would claim they had seen San Borodon, just as old fishermen of Galway still believe they catch sight of O'Brasil," the blessed isle west of the emerald one.

Welcome topped off at Santa Cruz de La Palma, taking on diesel fuel, water, frozen chickens from Uruguay, the fruits and breads of the local markets. Word in the harbor was that a big storm was working its way down the coast, and smaller boats were changing their departure plans. We called the National Hurricane Center in Miami, which tracks big disturbances from Africa westward. No storm, said a tropical meteorologist. The big high, its eastern edge on the Azores, would keep any trouble well north of us. Winds looked good, fresh northeast breezes from 10 degrees to 25 degrees north all the way across to the Caribbean: the trade winds, truly the breath of heaven to the first men, whoever they might have been, to have run before them across the North Atlantic to the Americas. But the trade that subsequently developed along that "southern route" is not the source of the name. In early days to "blow trade" meant to blow constantly in one direction.

We left La Palma, heading for Hierro, the southernmost of the Canaries. The afternoon sea carried the sky in facets, blues flickering purple, water so flat that when a lone whale spouted far astern we could see his fountain as clearly as if it had been a single bush in the desert. No matter; the engine would take us out from under the calm.

But it would not. After a few hours, Carrie Snyder, at the wheel, felt and heard a slight surge. Mike Jones and Willy Claxton investigated. The bolts holding the coupling of propeller shaft to transmission had sheared. We drifted down on Hierro as four men took turns on their bellies teasing the diseased bolts from their flanges. For the first three hours it looked as if Snyder were going to have to call for a tow. Three hours later new bolts were in place and locked with tin from an old can. Everybody was too busy to think of the ghost fleet crossing our wake in the early dark, the three small ships running west before a strong northeast wind. Snyder started his engine, and we motored off southwesterly to meet the trades.

We found them in a day, the northern edge, just about where the hurricane center had predicted they would be. The sky still didn't have a trade-wind signature — formations of small cotton clouds sailing over. Clouds were there, but they were large and flat-bottomed. And the wind, after a first fine push, died away. Mike chirped at the helm, whistling the breeze back from where it played, miles off to starboard. Howard Freeman said friends of his were becalmed on their crossing and came into Antigua three weeks late, with neither food nor water. Willy Claxton looked west out of his bright blue eyes and said something softly. My ear, at last attuned, translated it almost immediately: "Nothing out there for thousands of miles." And "Really puts us in our place." Suddenly, strangely, the word *calm* had become insidious.

Early or late, all of us were adjusting to the shock of being together, most of us still strangers, our place a contrivance of muttering wood forty by fifteen feet at best, moving across the interface of two voids. In the following days we became far more considerate of one another than if we had been of one blood. We strained to be pleasant, sensing dissolution if we fell to snarling. We ate as many meals as we could together to protect our gossamer unity. We asked one another "How did you sleep?" because we needed to know the states of alertness of our watchmates. When seasickness kept me abunk for an afternoon and night, I was cared for with tenderness and allowed to sleep through my assigned midwatch. Through the voyage, I paid back those who had stood in for me. Balance, in this small society, was all.

Gradually, as the trades strengthened, we relaxed in their regularity. We came to accept open water. The routine helped. Living on a boat is an exercise in housekeeping for sardines. No personal space, little of any other kind. Six people in *Welcome* are one too many for true comfort. Lockers, cupboards, drawers, every inch of stowage space was taken, some with gear I had brought along as insurance and never needed. We built routines on routines to balance constraint. A quart of fresh water a day for drinking. Salt water for dishwashing and some cooking. Keep neat. Keep clean. Wash your clothes in a bucket dipped in the sea. Take saltwater showers under the hose forward, and be quick to dry yourself or the sun will bake in the brine and shrink your hide.

Stand your watches, two with each rounding of the sun. Snyder ran the twelve-to-four, Carrie the four-to-eight, Mike the eight-to-twelve. We didn't dog the watch (divide the afternoon four-to-eight to break the routine). Instead, Snyder rigged a staggered watch so that no two people had to put up with each other for more than two hours of duty. Willy manned the two-to-six, Howard the six-to-ten, I the ten-to-two. One hand for the ship, one hand for yourself. After some days, *Welcome* seemed to impose the discipline. She became the repository for our responsibilities, and in this sense she lived.

The southern route from Europe to the Americas began in sail and remains in sail. Modern shipping lanes, free of the wind, run elsewhere. One day we saw a sail north of us, and then another. We raised an American voice on the VHF radio, a youngster and his friends taking a big yacht home through the Panama Canal to her owner in Oregon. We were not alone. "The North Atlantic may be likened to a giant roundabout," reads the blurb for a Baedeker for sailors of the gyre. "The ocean currents and the weather systems revolve clockwise about a center, located some distance west of the Azores. The intending transatlantic sailor has but to hop on the roundabout to be carried to his destination."

We had the Baedeker aboard, a manual called *The Atlantic Crossing Guide*, full of old-boy advice on navigating, weather, provisioning ("If you have the right sort of wife, forget everything and leave the whole job to her"). We had EPIRBs (Emergency Position-Indicating Radio Beacons), small units that broadcast special locational signals which can be picked up by aircraft or satellites and relayed to rescue stations. Mike, who brought his own, told me that the cricketing from one EPIRB, picked up and passed on by the Russians, had been tracked down to a loft in Glasgow. It had evidently been stolen, and the thief, in his haste, had jostled it awake.

We had not only the VHF radio, for short-distance communication, but also, mounted inconveniently on a leg of the chart table, a single-sideband radio, a complexity that bounced its waves off one of three layers in the ionosphere and reached out thousands of miles. Carrie was its mistress, and she could raise high-seas operators in static thick as gruel. We talked regularly to families ashore and, after each session, glowed with a security that Willy earnestly sought to dispel. "One minute you're talking to the world," he said with some scorn, "the next you're in a rubber raft fighting for your life, and nobody to help you." Snyder shared some of Claxton's resentment of the SSB. Having placed the thing where it could barely be operated, having admitted that it scared him, he asked Willy where he would have put the big radio. "About twenty feet aft," said the Cornishman. In the wake.

The ghostly presences to the north had no such symbols of security, but they did have about every size of luck during their thirty-three days of crossing. The fleet took its departure from the Canaries in mid-September by the modern calendar, a good time for hurricanes (the season usually ends around mid-November). Yet the weather smiled. The vessels sailed just a bit south of west for most of the outbound journey, touching the horse latitudes. Yet the Admiral was not seriously becalmed. "Jesus and Mary be with us on the way," he was fond of saying. It appears they were.

Welcome took a small chance in turning westward before getting south of 25 degrees north, but Snyder decided to test the hurricane center's forecast. If he missed the trades he could motor south into them. The wind held. Snyder too was a lucky captain — with a diesel.

Westward, then, *Welcome* and her spirit escorts, separated at most by a few hundred miles as the schooner drove through the days down to Antigua's latitude. The captain to the north wrote each day in his journal, but all we have are bits and paraphrasings, the work of Bartolomé de Las Casas, the first priest to be ordained in the newly conquered Caribbean. The Admiral, Las Casas says, wrote that the weather was like spring in Andalusia: "The sea was smooth as a river and the finest air in the world . . . Nothing was wanting but the singing of the nightingale."

So it was, for a time, with us. The moist air softened my aging skin. The sun, while our latitude was still high, was a delight. To the north the ghost fleet caught fish — tuna and dories. Willy rigged a line with a Finnish plug and dragged in his own dories, rainbows dying on their flanks. He fried them, and we feasted. Columbus and his people noted the swallowtails that came aboard in the night. We called them flying fish and ate them, too. Superb. A taste like smelt.

Our surroundings moved westward with us. Winds aloft varied somewhat, but in the main they veered slowly as we went along, from northeasterly to easterly. Under them, the sea made its own crossing. The Canary Current, which was pushing under our starboard quarter on the run down from Portugal to Madeira, setting us down or southerly off our course, had turned with us and now nudged the stern, adding something like half a knot to our speed. Not much in a day, but in a full crossing this new current, the North Equatorial, might easily account for 220 out of the 2,600 miles traveled — the equivalent of a day and a half's run at *Welcome*'s speed.

The ghosts outsailed us on most days. They made up to 180 nautical miles dawn to dawn, and our best run was 157. That had something to do with their size. *Niña*, the Admiral's little girl, had a length of sixty-seven feet by current accounts; *Welcome*, counting her bowsprit, measures sixty. What counts most in speed, though, is length at the water line. Bluff-bowed *Niña* probably measured around fifty feet there against our thirty-six, and that more than offsets the difference in sailing efficiency of the two rigs. *Welcome* never got much over six knots, except in squalls. I reach that speed jogging along beside the Deerfield River. But six knots in a sailing vessel gives a sense of power that cannot be found on land, a power produced by a sly offering of our cloth to the brute wind, by a translation of strain into motion. It was a suitable pace for the schooner. More would mean, eventually, exhaustion of and in the boat.

The Atlantic Crossing Guide speaks of "the glorious downhill ride through the trades," and the adman tone doesn't diminish the accuracy. The days beginning and ending the crossing were full of wind and following seas running more than twelve feet. We climbed and dipped, the counter on the taffrail log lagging and then spinning. All around, waves played against great swells coming in at an angle from some storm hundreds of miles away. Snyder, gazing out on the welter, pronounced it a force six on the Beaufort scale, in which zero is flatass calm and twelve is hurricane. That put us in the strong breeze category, twenty-two to twenty-seven knots.

The independent swell and the confused seas it brings are typical of the trade belt. As Columbus slipped south into it, two things happened of benefit to his cause. First, the fickle weather of the horse latitudes produced a head wind. Instead of complaining about being held up yet again, the crews rejoiced. See, they said, we can get back to Spain after all. And then, the Admiral wrote, the sea rose without wind, which caused great astonishment on deck. "The rising of the sea," added that modest man, "was very favorable to me, as it happened formerly to Moses when he led the Jews out of Egypt."

My introduction to handling *Welcome* in the rough came not far west of the Canaries. A fair breeze had sprung up in the evening, coming in over the starboard quarter and making for a broad reach. We sailed into the night with a lot of canvas. Just as I came on watch and took the helm, a squall hit us and the wind jumped to thirty knots in less than a minute. *Welcome* heeled and ran up into the wind while Mike and I held the helm full against her. For a short time we hung on the wheel, helpless, watching the sea kiss the port rail.

Snyder and Claxton, the skipper in his drawers, tumbled up the companionway. In less than two minutes they had the main down — that was the demon, that great sail was powering the turn — and *Welcome* returned to her senses. Afterward we jollied each other, gentling what remained of the tension. *Welcome* had done the same thing off Finisterre, when Art and Carrie had run into sudden wind. The ship lay down under it, and the rudder was so far out of the water that it was useless. They started the engine and ran out of harm's way.

We learned from the squalls that the square sails were the crossing sails, and day after day the forecourse and topsail pulled us across the sea. On the southern crossing you roll half a million times, give or take. *Welcome* did a complete roll every four seconds or so, and the inclinometer over the chart table normally registered tippings of well over twenty degrees. Added to pitch and a bit of yaw, the roll produced the opposite of what astronauts experience. We were subject, constantly, to weightfulness. The G's waited — and weighted — everywhere. You wish to go aft from the forward cabin. You start out on course, slam into the gimbaled stove, pivot just enough to catch the leading edge of both hipbones on the saloon table, and jackknife. Weightfulness can catch you brushing your teeth in the after head, a vertical coffin of a place, and throw you out the door, breaking its wooden latch, and onto the exceedingly solid butt of the mainmast. A champion performance consists of keeping your meal on your plate.

But adjustment comes, or something passing for it. After hours at the helm I learned to steer well, fighting my tendency to counter *Welcome*'s every move and letting her find her way in that rough country. I was at the wheel one glorious night when the squalls sailed in astern one after another, one a great manta winging in under the moon, and the schooner fought to come up and duel with the wind. I straddled the steering-gear housing and kept her in check, exulting, feeling how she rolled up in the water rather than down as some boats do. When the skipper took over the watch, I asked him if I could marry his boat.

Snyder, still sleepy, started and then giggled and shook his head.

Well then, could he put a western saddle on the goddam housing to ease the bad news for the butt?

He came awake, looked at me out of a hard face inches from mine, and shouted slowly, "No way! You're just the son of a bitch who would wear spurs and ruin the brightwork." Then he fired off a laugh that could call up a gale and punched my arm, and we settled down to sail.

My new skill freed the more experienced crew to adjust and repair the rig, to keep the deck clear of what Willy called "gubbins," a wonderful word, Brit for gismo. *Welcome*'s light square sails were never meant for the constant stress of an oceanic crossing, and their rigging wore and broke. Going aloft in the roll was sometimes impossible, always difficult. I spent more and more time behind the wheel as we experimented with courses that would minimize movement.

Willy went up, his harness clipped to a halyard, and then Mike. The cook, sure of a strength he had whetted with white-water kayaking in the Scottish highlands, knitted himself into the rigging and sweated as he rove rope and jury-rigged fittings Snyder had put together. I sat on my hard nonsaddle and thought of an old Cornish fisherman reminiscing on a tape recording Willy Claxton had given me. When you come back from sea, he said, oh, that's fine. But when you're outward bound again, "you're on the knucklebones of your ass." Truly.

The days ran with the schooner, and we paid great attention to the small decisions, enlarging them against the settling routine. Today I shall heat a cup of my water ration and shave. Today I shall wash my drawers. Today I shall not take a hose shower unless the sea slackens. Books helped. They come from forests, and their smell is a land smell. I read in them about our ghosts to the north, and their days became mine. I sat in the saloon eating chicken tandoori and thought of their sour wine, their biscuit and cheese going bad, the water foul enough to gag on. Yet, the Admiral wrote, only one man took sick, for a few days, with the stone, and he old and accustomed to the complaint.

Our days and theirs went by the watches. They had the advantage in routine, the discipline of religion. Ship's boys sang as they turned the *ampolletas*, the sand glasses that measured the half hours, as they called the crew to the several services: tierce, vespers, compline, the prayers at the setting of each watch. Morison writes at his best about the gentleness and complexity of these observances, these ceremonies amid the stink of bilges, the constant thrashing of the tubby hulls. Daybreak. A young ship's boy salutes it with, "God give us good days, good voyage, good passage to the ship, sir captain and sir master and good company, so let there be, let there be a good voyage, many good days may God grant your graces, gentlemen of the afterguard and gentlemen forward."

On *Welcome* we prayed alone, each to his own brand of god, each in his own times. In the long stretches between, we tried a little levity. Someone said that when God made man She was only joking. Howard Freeman had brought a Bible with him, wrapped in a Baggie to keep out the salt. It stayed in his duffle mostly. But then a sea would bully in over the counter, or we would call for a shipmate up forward taking the sun and, for two seconds too long, get no answer. A plea then, silent, repeated: Not yet, not now, please. Sometimes after those small scares the Bible would appear, and someone would read the poetry of Ecclesiastes: "All the rivers run into the sea and the sea is not full; unto the place whence the rivers come, thither they return again." And, "The wind goeth toward the south and turneth about toward the north; it whirleth about continually, and the wind returneth again according to his circuits." I think we took a certain strange comfort in our insignificance as *Welcome* moved along the great whirl of the gyre.

Eating gave us comfort, too, and the talk in the cockpit. Carrie and Art told stories from the northern crossing, when they took *Welcome* from Boston to Scotland. They ran into icebergs floating down from the Labrador Sea. Well, Willy Claxton said, an old sailor told him one time in Falmouth about the bergs. "They smell like cucumbers," he had said.

Squalls stalked us some nights, with thunder and jetting rain — nothing serious enough to block the companionway with washboards or take down the canvas hood — the dodger — to keep it from being ripped off in a tangle by green water. (We figured if that happened when the skipper was coming on watch, we might end up with an Artful dodger.) The Admiral, sailing among his imagined islands, had two false landfalls, when over-eager lookouts confused clouds on the horizon with what they strained to see. We, seeing those same clouds, could understand. And one night, Mike and I smelled something in the wind. I thought it was bacon frying. Mike said it was the smell of the Sahara a thousand miles astern, brought to us by the winds of that desert that have been known to carry dust to the Antilles and Florida. Perhaps the Spaniards too smelled food cooking, tasted grit in their mouths, and wondered which of their demons was tormenting them.

We sailed mostly under gauzy gubbins of clouds, and worried more over winds that were dying than rising. The whole ocean would ease then, flashing full of sun, in a mood perhaps like the one an oceanographer aboard the space shuttle looked down at, looked again, and grabbed for his camera to record. What he saw was a pattern of whorls on the skin of the sea that no one had seen before. He photographed them when the angle of sun brought them out like grain in good veneer — currents plaiting with currents, only some of it understandable. Something over halfway across, we began using the engine for hours, and every time the diesel sounded we checked remaining fuel against remaining miles. Then the trades breezed up, and Snyder, regarding the cloth he had just ordered aloft, would say, "The engine's great when it's on, but it's better when it's off."

We all lost weight to the isometrics of roll and pitch. Snyder proudly pulled out the waist of his pants and grinned at the space he no longer occupied. He kept his impossible pace, worrying, punishing himself for his mistakes. He went at things in a slap-dash way, but some part of his mind followed along like a hound and pointed to the undone and the errors. In returning to them, he discovered other jobs for himself. The net effect was prodigious and, in a jury-rigged way, precise. We tried to get him to wear a safety harness in the hard squalls. We worried that he would let fatigue trip him. He merely stumbled now and then, and Carrie or someone else would holler at him, but he paid no mind. He'd be up forward with the young Britons tuning the rigging, and you'd hear from behind a billow of main course a mad voice yelling, "I thought I heard the captain say, 'One more turn and then belay!' "

I wondered about the unknowable to the north. "Xpo FERENS," he signed himself; the Christ-bearer, Saint Christopher, who carried the Child across the torrent. It was he who would bear his namesake to Cipangu and beyond, and he begged his sovereigns to invest the fruits of his "Enterprise of the Indies" in the conquest of Jerusalem. If a portrait were made of him during his life, it does not survive. In some later representations he looks to me suspiciously like George Washington. He was a Genoese among Spaniards, a man of humble birth among grandees, and a captain among captains far more experienced in command. I thought of him, perched perhaps on the board slung over the ship's side for purposes of relief — the *jardín*, they called it, the garden. There, hidden by the night, the Admiral must have had thoughts that only the most obstinate of faiths could have tamed.

He sailed as we sailed, in winds faithful as lovers. He saw the life we saw, like the white tropicbird with its streamer of a tail and its affection for ships. He saw what we didn't — high flocks and low straggles of migrating sparrows and warblers and thrushes crossing the Sargasso, some of which "came to the ship singing." Like us, he steered by Venus, bright over the bows. Unlike us, he steered toward a figment. Yet in his ignorance and self-deception, he was wise enough to deceive his crew, placating them with reports of distances sailed that to him represented a reduction of almost ten percent. But his own computations of speed, gathered from his primitive chip log or observations of spume or weed floating by, turned out to be excessive in just about the same amount. The crew got the truth and, in time, sailing away from everything they knew, the truth came to hurt. Mutiny rose to likelihood in the last days of the voyage.

The Admiral, like most European mariners of the day, kept counsel with the North Star. He relied on the constancy of its position, and using his eye and a measuring device called a nocturnal to ascertain its height above the horizon, he kept well to the latitudes he had selected. So important to navigation was Polaris that when the *marinheiros* coasting Africa reached latitudes below 10 degrees north, they feared to go on. At that point their holy star was too low on the horizon to be of use, and they feared what would happen to them if they lost it — far more than they feared the boiling seas and the people who walked upside-down and the monsters beyond their bows.

Morison thinks that Columbus carried a quadrant with him but that he thought little of it. Like the marine astrolabe, the quadrant was apt to be a rudimentary conversion from instruments designed for land surveying. Aboard ship it was almost impossible to use, with its peephole and plumb bob. When Columbus took it ashore in the Caribbean for a sight, he got a latitude reading that put him just about at the height of Cape Cod.

It was in navigation that the centuries got the better of the few hundred miles separating us. The ghost fleet traveled with charts of the Atlantic that were crafted in the main out of pure imagination. *Welcome* carried a drawer full of large papers giving us locations, depths, hazards, lights, radio beacon frequencies, and prevailing winds and currents by the month.

Snyder is, I think, a priest of the sun before he is a Quaker. At least twice a day he went topside with his sextant to sit, facing his deity as it arced or hung in its nooning. He would adjust the filters on the instrument to protect his eye, then make his sight. The idea is to get sun and horizon in proper perspective and then, tilting the sextant slowly, like a pendulum, bring the image of the sun's bottom edge down until it touches the image of the horizon. At that precise instant Snyder would sing out "Mark!" and his acolyte would jot down the time of sight completion to the second, followed by the sight angle, in degrees, minutes, and seconds.

That is the miracle of location: time, reliably measurable time. *Welcome* carries a *History of the Navigator's Sextant*, and it says:

> Of all the techniques for position-finding at sea, including a vast array of sophisticated radio and electronic aids, nautical astronomy continues to provide the most reliable technique of all. Provided that the mariner is armed with a Nautical Almanac, means to compute his nautical and astronomical problems, a reliable timekeeper [a wristwatch] and an efficient means for measuring altitudes [a sextant], he is self-contained as far as finding his ship's position is concerned.

With precise time you have precise longitude. That is why the British for so long offered a handsome reward to anyone who could come up with an accurate ship's chronometer. Parliament announced the prize in 1714, set it at £20,000, and then had to wait until 1761, when a device developed by John Harrison was tested on a course to Jamaica that lay close to our own and was found to lose only five seconds in nine weeks. Good, but the wristwatches aboard *Welcome* did about as well — and we had a radio signal against which to check them.

In his navigating Snyder deviated not a degree from true character. He took his numbers down to the chart table, made mistakes, swore at himself, played with his calculator (making *plunks* with his mouth as he hit the keys), labored, hunted down his errors, and came out with positions that, if they had been bullet holes in a target, would have made a master marksman proud. He worked so hard at repeated positionings, he said, because of what it might mean in an emergency: "If you're in trouble and you send out location by whole degrees, a plane has about thirty-six hundred square miles of ocean to search." His way, the target area shrinks to four to six square miles: still a lot, but a lot less.

None of this for the Admiral. His *ampolletas* gave him some idea of time, good enough for the ships' routines and little else. And he could check them against the movement of certain stars around Polaris. But his reliance in the passage had to be on compass, dividers, a lead line for navigation on soundings, a poor excuse for a sea chart — and meticulous observations, meticulously recorded, of the movements of wind and water. Here he was superb. His errors of overestimation were constant errors, not random ones. He could not measure currents accurately out of sight of land — no one could then — but he did notice their effects. His sense of the wind was exquisite. I can see him, feeling for it with his hands as we did, sensing it on the back of his neck as we did. An error in course of 6 degrees on the southern route would mean an error of 250 miles at its ending, and Columbus would not tolerate that. He writes in his journal that when he left the Canaries the seamen were steering badly. (I extend lubberly sympathy to them.) There was only that one mention. Something, a rope's end perhaps, thereafter convinced the helmsmen to keep a straight wake. To Columbus an error of ten leagues in a thousand was a poor show. "No such dead-reckoning navigators," writes Morison, "exist today."

We were closing on our destination now, and the first twinges of channel fever began, the raging for home. The petty annoyances of weeks of enforced companionship became harder to dispel, our minds chivvied by time flowing past at six knots. If we make 140 nautical miles a day, well then, in three days — and we scampered forward, past the bow and into next week.

In the last days the wind rose and drove us, and in the night dolphins herded us along. We could hear them going *poo* as they blew alongside, and I saw one thrusting head burn for an instant in green, living fire. Mike Jones told me that when he had sailed to Brazil some years back, he had installed stereo speakers in the cockpit; the dolphins loved the music, he said, and they stayed around longest for drums and acid rock.

We sailed through the shipping lanes between North America and Rio and Capetown. A big tanker came up on our port quarter when I was at the wheel, and I got a lesson in steering a true course. She came on fast, about eighteen knots to our six, and didn't alter course until Mike shined a powerful searchlight on our sails. I kept looking astern and *Welcome* wobbled accordingly until Mike rapped me on the elbow. "You steer two-eight-eight true," he growled. "Do you want that bloke to guess wrong about that?" I kept eyes and mind on the compass lubber line and so heard but did not see the monster slide by a scant quarter-mile down our wake.

We had seen little of man all the way — a fishing float here, a forty-foot segment of gill net there. The tanker put us back in the world. We heard the signal from Antigua on the Radio Direction Finder. Late on the nineteenth day out of Santa Cruz, we raised the island itself, a smudge of green-gray light through the squalls.

The Admiral had better luck in landfall than did *Welcome*. As he neared the Bahamas, scudding before trade winds blowing a near gale, his crews saw floating logs, and one man fished out a carved stick. They heard birds passing all night, and Columbus, says his chronicler Las Casas, "knew that the Portuguese had discovered most of the islands they possessed by attending to the flights of birds." On the midwatch of October 11–12, the lookout on the *Pinta*, a sailor named Rodrigo de Triana, shouted "Tierra! Tierra!" and the dream of discovery became discovery itself.

Morison has the Admiral sailing over to the *Pinta* and hailing the captain: "Señor Martín Alonso [Pinzón], you have found land."

"Sir," replied that often rebellious officer, "my reward is not lost."

"I give you 5,000 maravedis as a present." That's a few hundred dollars at best, assuming Columbus was referring to the copper coin of the time — a bargain for a sighting that would go a long way toward rescuing Spain and all Europe, for a time, from its depression.

A few hours before he made land, Columbus thought he saw a light, like a little candle, rising and falling. It could have been a fire on one of the small islands before him in the night, or someone jacking fish in the shallows. Or it could have been the strain of passage and command tricking his eyes. Whatever it was, it astonished him. We sailed 2,600 miles toward the light outside English Harbor in Antigua, supposedly one of the easier landfalls in the Caribbean, only to find, after hours of peerings and halloos and false discoveries, that there was no light. It had gone out months before, and no one had fixed it.

We were more than astonished. Coming in to the land is among the most dangerous of nautical undertakings. As he explored among the islands, Columbus got careless about relying on his sounding line and so lost his flagship to the shallows. We coasted along well out from Antigua but on a course that would bring us close to a reef if we were to miss English Harbor entirely — which we did, as we finally discovered, and turned back into a pounding sea. With the diesel hollering at full bore, we crept at a knot or two toward the land. After a couple of hours we picked up the red range lights by the dogleg of the harbor's entrance. The wind gave up on us then, and the seas, and we puttered in to anchorage and — at three-thirty in the morning — to sleep.

The lick of slack water kept waking me up in the sunlight. I could not get used to it, or to the sound of voices from the yachts anchored near. I peered out at the high, arid hills of southern Antigua and saw the British battlements where my brother and I, thirty years before, had dug up bottles and buckles dating from the time Lord Nelson and his fleet fought the French in these parts. On this familiar ground I thought of Columbus, enchanted as the sun rose on his totally unfamiliar world. He remained so for weeks, to the consternation of some in the crew. He told them that a thousand tongues would be insufficient to inform the king and queen of what he saw there, and he wrote his sovereigns that "the singing of birds is such that it appears a man would wish never to leave here," and "I assure your Highnesses that these lands are the most fertile, temperate, level and beautiful countries in the world." Not just self-service, I think. Muna Lee, a poet intimately connected with Latin America, wrote in the 1940s that something happened to those who came to conquer. She talked of a conquistador describing the big green fireflies of Santo Domingo so well that "they actually flutter across the page." That was entirely new, she said: "Spanish prose and poetry are peculiarly lacking in the visual perception of nature that is so integral a part of English literature — except that prose and poetry written in Spanish but dealing with America or produced in America." Clearly, the American muse made no distinction between Genoese and Spaniard.

The new, of course, never remains. What the Admiral discovered is largely gone. The animals he brought with him on his voyages went wild and multiplied and trampled the ground cover. The seeds that came with him and those that followed seized the torn earth and made it their own, so that much of what we now see in this hemisphere is biota of European origin. The viruses and bacteria of the newcomers decimated those who had run to the shore, staring at their first sails and shouting, "Come and see the celestial people!" These were native populations, isolate populations with no defenses against what lay beneath the sails, and the scourge was in some cases far worse than the Black Death had been in Europe.

The invaders themselves may have marveled at what they saw, but they coveted more what they could make of those sights through slavery and pillage. The same man who wrote his sovereigns of beauty beyond compare spoke also to his god: "Our Lord in His Mercy, direct me where I may find the gold mine."

The lassitude of English Harbor in the fifties was also gone. The dockyard buildings had been reconstructed, turned into hotels and boutiques. The harbor was jammed with expensive yachts, charter boats in to show off their machines and cuisines to visiting travel agents: *Darius* from Wiesbaden, *New Wave* from Helsinki, *Mr. Bullfrog* from San Francisco. I had seen boats like them in every port of the passage, the new, monied migrants of the Atlantic. Round and round the subtropical gyre the Euros chase the Yankees. A couple of thousand miles back down *Welcome*'s track, three hundred vessels were running before the wind, bound for Barbados. Three hundred, part of a cruise-race that left the Canaries in one great blotch of plastic and Dacron. Alone, alone, all, all alone; alone on a wide, wide sea.

I left *Welcome* that morning. She would go on, up the line to St. Martin and then, the coming May, north to Bermuda and home. Mike and Willy took me over to the dockyard in the rubber dinghy, and I began making my adjustments, shifting energy from experience to memory. I had learned enough of sailing to advance, perhaps, from movable ballast to an elderly ship's boy. I had learned about gubbins. I had learned that flying fish taste like smelt and icebergs smell like cucumbers. I had learned to be at great ease with an anomalous Quaker. When I left Art Snyder, he told me he had liked having me aboard, and I told him I had liked being aboard, and then, looking straight ahead, we both put an arm around each other and took a brief, hard purchase.

Out beyond the mouth of the harbor, the trades still boomed and the sea blew by into the Caribbean, on its way around its circle. I went to a hotel to sleep, and in the afternoon awoke to an alto voice singing in the street below "O come let us adore him." The continuum of voyaging dropped away, and the calendar returned. I went out to the boutiques to shop for Christmas.

4
Following On

"AT A VERY TENDER AGE," Columbus wrote, "I entered upon the sea sailing." Most of his failures came when he left his ship to act the conqueror, the administrator of his cherished Enterprise of the Indies. His greatest discoveries remained not islands in the tropic sea but routes to reach them. Europeans who followed him extended those routes, and in that process the Gulf Stream system came to influence, to a remarkable degree, the settlement of colonies and the patterns of their commerce.

The Spanish took hold in the Admiral's islands a full decade after the first landfall, when one Nicolás de Ovando arrived in Santo Domingo to govern for the Crown. Ovando learned his governing in what the Spanish called their *reconquista*, the ousting of the Moors from Spain by Castilians and others, the Iberian invasions of northern Africa, and, most important, the subjugating of the Guanches in the Canary Islands. There Spain learned the principles it would follow in the New World: conquer, build a colonial society on that of the conquered or, if the natives cannot bear up under slavery, import peoples who can. In the New World the imports were blacks, supplied initially by the Portuguese, who had learned the trade in their own probings and pillagings along the African coast. "Wherever the European hath trod," wrote Charles Darwin aboard the *Beagle*, "death seems to pursue the aboriginal."

The pace and verve of those Spanish astound me. Fifteen years after Ovando took over the Enterprise, the conquistadors were coasting in the Gulf of Mexico. Fifty years later, they had taken the Aztec and Inca empires. In one century they came to control the land from what is now the western United States to southern Argentina. They were as predatory as most other European societies, but there was a method to their predation that turned enterprise to empire.

Contrast the vigor of the Spanish outreach with that of their Atlantic competitors. Five years separate Columbus's virgin voyage from the explorations of his fellow Genoese, John Cabot, down the northern coasts of the new continent. Yet one hundred years separate the settling of Santo Domingo and the English attempts to found Jamestown. Most of the French efforts in that interim, and some of the English, went into fishing on the great banks Cabot had crossed. The French fur-trading post at Quebec was established in the late year of 1608.

Columbus elected for the southern route because, according to his charts, it lay close to the latitude of Cipangu, and because, according to his hunch, the northeasterlies around the Canaries blew across the ocean. He selected his first homeward track in the same manner, betting that if he beat northward through the trades he would eventually find westerlies to blow him to the Azores and Spain. The winter of 1492–93 was a terrible one in Europe. Ships drowned by the hundreds, and a storm almost killed the Admiral and his *Niña*. He was forced to run into the Tagus, and there took the sweet vengeance of telling the mighty of Lisbon that he had found for the Spanish what the Portuguese had earlier refused to let him find for them.

As the conquistadors moved west through the Caribbean, Columbus's home route became increasingly inconvenient. It took longer to reach the windward passages to the Atlantic, and political rivalries along the island chain could endanger a ship as much as the reefs outside the harbor mouths. An alternative was needed to maintain the momentum of conquest. It was found by an *hidalgo* who had signed on, at the age of nineteen, for Columbus's second voyage. A redhead like his commander, Juan Ponce de León was ferocious in battle and ambitious in what passed for peace. He conquered Puerto Rico, governed it, enriched himself, grew bored, and set off to find new lands. One in particular interested him, the land of a certain spring that relieved what was known at the time as *enflaquecimiento del sexo*, a waning of amatory power. Why a man in his late thirties would need to take those waters is not known. Perhaps he was anxious to please his elders.

The story has a uniquely American flavor to it: Ponce de León coursing around Florida a bit less than five centuries ago in search of the fountain of youth; and today, the rich and the elderly of the continent flocking to the same place for just about the same reason. But the great find of that Easter went generally unnoticed. Antonio de Herrera y Tordesillas, a sixteenth-century Spanish historian whose accounts are accepted as reliable, quotes from a journal of Ponce de León's voyage:

> Sailing south and entering . . . the broad waters of the sea, all the three ships on the following day [April 22, 1513] saw a current which, although they had a good wind they could not stem. It seemed that they advanced well, but they soon recognized that they on the contrary were driven back, and that the current was more powerful than the wind. Two ships, which were somewhat nearer towards the coast, came at anchor, but the currents were so strong that they made the cables tremble. . . . The third vessel, which was a brig, being a little more towards the sea, could find no bottom. She was carried away by the current, and they lost sight of her, though it was a clear and calm day.

Ponce de León coasted down Florida in the southward-flowing waters inshore of the current he had just discovered. He went on to sail across the Yucatán Channel and into the Gulf of Mexico. On his second voyage, which touched on the west coast of Florida near what is now Fort Myers, he took an arrow from a native bow. The wound festered and killed him.

For my purposes, the hero of the tale was not the conquistador but his pilot. Antón de Alaminos was also with Columbus on the second voyage. If the Admiral had a wind rose in his head, Alaminos had a rutter in his. He remembered everything of the seas he traveled, including the Florida Current, and he put his memory to shrewd use. Increasingly, when an expedition was being mounted to move west and north into the Gulf of Mexico, the call would go out for Alaminos.

In 1519 the call came from Hernando Cortés, subjugator of the Aztecs. There was gold from the great city of Tenochtitlán to delight the king of Spain, and Alaminos carried it. His instructions were to make the fastest possible crossing and to stay away from sea lanes controlled by Cortés's many enemies. He did so, remembering in the rutter of his mind the set of currents in the Gulf of Mexico, letting them help him around to western Florida. When he passed Tortugas and the keys, he took an enormous gamble. He remembered that brig being carried away by the sea that clear day at Eastertide. "He thought," says Herrera, "that those mighty currents ought to empty somewhere into an open space."

The open space for Alaminos lay north of the Bahama Banks. He evidently passed fairly close to a sprinkling of islands that after the arrival by shipwreck of one Juan de Bermúdez in 1526 became known as Bermuda. (It is said that the king of Spain was delighted with Bermúdez's news, since that gave him the opportunity to drain the islands' swamps, which were thought to cause the storms that made the place an unhappy one for mariners.)

The Spanish, like other seafarers of the day, kept Alaminos's new route very much to themselves for as long as they could. Their captains were told to weight their sailing directions with lead and to toss them overboard if capture were imminent. But the word got out in time. Richard Hakluyt, the great chronicler of British sea exploits, printed numbers of "ruttiers" concerning the courses the Spaniards kept from Havana to Spain. In summer, he said, you can go north of Bermuda and head for the island of Flores in the Azores and then on in. But, he cautioned:

> I advise thee, if in the Winter time thou be shot out of the narrowest of the Channel of Bahama, and would goe for Spain, that thou must

> goe East Northeast, untill thou be in 30. degrees [north] rather less than more; and then thou mayest goe East and by South, because of the variation of the Compasse. And stirring [steering] hence East Southeast, thou shalt goe to the Southside of Bermuda: and must goe with great care, because many have bene lost heere about this Island, because of their negligence. And when thou art sure thou art past this island, then goe East Northeast until thou be in the height of seven and thirty degrees: which is in the height of the island of St. Marie. And going thus, and not seeing land but seeing the Sea to breake, make account it is the rocks called Las Hormigas. Then to Faial [in the Azores] And then East and by South to Cape St. Vincent.

And there you have it, coming and going. Quite literally, Antón de Alaminos changed the course of Spanish empire. The galleons could have saved even more time if they had been directed farther north in the Stream before cutting eastward; that would have kept them clear of dangerous Bahamian shoals and put them on the great circle route to Cádiz. But the pilot's gamble paid dividends for two centuries, and not only to Spain.

A nineteenth-century German geographer named Johann Georg Kohl, to whose history of Gulf Stream explorations I am much indebted, was quite sure the Stream and its gyre had "influenced the . . . growth and progress of the entire human species." My fancy, in this regard, takes a lesser leap: I like to think of the Stream system as the organizer of the Great In-Between, the waters separating Europe and the Americas. Once the Spanish had established their sweep around the gyre, the British and French and Dutch began their variations on the theme. The French, for example, came along the southern route to burn Havana and establish outposts here and there along the southeastern coast, only to have them razed by Admiral Pedro Menéndez (who in 1565 succeeded in stemming the Stream from St. Augustine to Havana, something unheard of at the time). Spanish forts appeared as far north as the Chesapeake, built principally to protect the home route from enemy raiders. Undaunted, the English under Raleigh and his colleagues took and held ground in Virginia, which the Spanish considered part of their Florida.

In establishing the northern route, Cabot and his successors sailed a bit above the Stream to the Grand Banks and followed the Labrador Current down the coast. The distance was shorter, and the Banks provided a convenient provisioning station; word in Europe was that Cabot had found so many cod there that they could be taken "not only with the net but in baskets let down with a stone." Some British ships shifted farther south after 1602, when Bartholomew Gosnold came straight across the Atlantic — bucking the Gulf Stream extension a good part of the way — to poke around in rich waters along a peninsula he called Cape Cod. The Pilgrims started along Gosnold's route, aiming to settle in the Virginia lands, but not much is known of their passage. One account has it that they tried to cross the Gulf Stream and got shunted up to Cape Cod. That sort of thing did happen. Bartholomew Gilbert, coming up the southern route, stayed in the Stream too long, overshot the Chesapeake, and finally got off at New York.

Using the southern route to go much farther north than Virginia didn't make sense to the British. They divided their holdings, the southern to be serviced by the old route, and the northern — New England — by routes approximating Gosnold's. The Dutch in New Amsterdam (or New Netherlands, as the colony in what is now New York was called) stuck by the southern route, so the end result was a geographic anomaly: colonies less than two hundred miles apart depended on sea lanes that differed in length by three thousand miles and in latitude by as much as thirty degrees. The dividing line was Nantucket. The first landfalls on either route were generally settled first, and those in the middle of the Atlantic coast were developed later.

Once the colonies had goods to export and ships to carry them, they too sailed the gyre. In Boston and Philadelphia and, later, New York, the great trade fortunes began to accumulate, protected by the British navy. Raw goods went north and east "down" to Europe, and finished goods came back. That was fine. What was difficult was the coastal trade essential to prosperity. On southerly runs the Gulf Stream was a great hazard, forcing ships in toward the rocks and bars. Where it squeezed tightest, the cemeteries of the Atlantic occurred: Frying Pan Shoals, the shallows off Cape Lookout, and the killer, Diamond Shoals. You could go outside the Stream, as modern vessels like *Wilmington* do, and follow the countercurrents south, but then you often had contrary winds that could limit the mobility of a sailing vessel passing close to the Bahama Banks. In the Civil War, the Union Navy, out to blockade the South, ran many a ship aground. The same hazards are with us today.

It makes sense to me that the Caribbean would be important for Yankee traders, because their forebears had passed through it on the way to settlement. They knew the islands and the trade that could be had. The Spanish there needed few luxuries, but they did need barrel staves and the cheapest grades of codfish for their laborers. In the mid-seventeenth century, a Dutchman had planted sugar cane in Barbados, and the crop spread through the islands. Molasses was a by-product of the industry. Brought back up the Stream to New England, it made rum. Rum bought furs in the west and slaves on the African coast. Slavers carried their fellow humans in chains along the Equatorial Current to the Caribbean and sold them at high profits.

My great-great-grandfather, Moses Hillard, did not hold with slaving, but in other ways he was typical of the New England sailing captains. He was expected by his owners to be a sharp sailor and a sharp trader:

> The Brig *Sussex* under your Command, being now Loaded & ready for Sea, you will proceed with all possible Dispatch for La Guira [probably LaGuaira, a Venezuelan port] and there dispose of your cargo on the best Terms the market will admit and invest the proceeds in such Articles as you may judge most for Our Interest & return direct for this Place — In Case you are not permitted an Entry

at La Guira, You will if You think it advisable try one Other Port on the main, and also St. Thomas if it becomes necessary.

With that, and with a short list of going prices on the main (coffee, thirty cents; hides, eleven to twelve cents; indigo "one and three quarters to two and a quarter dollars") my great-great-grandfather, age twenty-five, put to sea. It was 1805, and the port of New York was coming into its prime as a transshipper to Europe of southern cotton, as a center of American trade on the Stream.

Moses Hillard knew firsthand about freebooting, the gray and often grisly practices that coexisted and sometimes melded with legitimate trade. In 1800 his captain lost the brig *Caroline* to a French privateer of four guns and fifty men while sailing home from St. Lucia in the Windward Islands. The French took most of their captives' clothes and the ship's goods, carried them to Guadeloupe, where Moses spent a month as a prisoner, "toughing it out in the usual way, half-starved" and working on French ships and their prizes in the harbor "to keep alive," before being sent to freedom on nearby St. Kitts.

When was a privateer a pirate? There were certain rules during wartime; you needed a commission to capture prizes then. But the distinctions broke down quickly, particularly in the Caribbean. The English sea hawks gorged their holds on Spanish wealth and sailed home up the Stream. The French, the Dutch, anyone with a strong back and a taste for blood could and did enter upon one or another adventure in sea robbery. Let us remember that Captain William Kidd lived quite well on both sides of the law, being a master at making friends in high places. Let us remember that Henry Morgan, one of the more violently inclined of the brotherhood, once served as lieutenant governor of Jamaica.

Smuggling began with the earliest shipments back to Spain. Much of the gold and jewelry that has been recovered from wrecked galleons was not on their manifests. It had been spirited aboard by agents for local gentry anxious to build up their accounts at home. Rest assured that of the four billion pesos of precious metals mined in Latin America during the Spanish empire (about nine tenths of it silver, the rest gold), a very large amount never saw the inside of the king's treasury.

As trade developed, ships nipped into ports from which they were banned by law or picked up cargoes their masters knew they should not be carrying. Smuggling became so common that customs officers took to saying that payment from a merchant was surer than payment from the king. In Boston an enterprising soul could take out insurance against being caught. It took the Civil War to stop the smuggling of slaves. It took the repeal of Prohibition to stop or at least slow down the smuggling of liquor.

In our own time, the Gulf Stream system continues to abet smugglers. Not long ago the Reagan administration declared a war on drug smuggling — and then appeared to forget about it. Cocaine and marijuana coming into the United States, a good deal of it up the Stream, now rank among our most valued imports. Early in the eighties, before cocaine became so popular among trendy Americans, marijuana from abroad was so popular that it made of the United States a net agricultural *importer*.

In the spring of 1982 I spent a week on patrol with a vessel involved in the drug interdiction program. She was the medium-endurance cutter *Dauntless*, 210 feet long, dressed in regulation white with the regulation orange-red racing stripes down the flare of her bows. When I went aboard she was the best drug buster in the service. Her record showed on her stack: forty-eight stylized *cannabis* plants painted in neat rows representing forty-eight vessels — sloops, coastal freighters, shrimpers — carrying more than a million pounds of marijuana, all told.

Because of its bulk, most of the marijuana came from Colombia by sea, along with cocaine, Quaaludes, parrots, guns, and other proscribed merchandise. Then as now, the cash flow in Miami from this trade was unbelievable, and across the state drug millionaires flourished. I was told that so much money was entering the system in Dade County that the Federal Reserve office there was kept busy shipping greenbacks off to other districts to preserve order. The magic of the market.

The clouded inshore waters began to clear as we went south along the Florida Keys, and the Stream showed dark far to seaward. Toward evening the skipper, Commander Mike Murtagh, a stocky man with a heavy and handsome face, invited me down to his quarters for a briefing. We were heading, he said, for the Yucatán Pass, the hundred-mile-wide cut between Mexico and Cuba. Smugglers loading along the Colombian coast could approach the United States through the Yucatán channel or through the cuts the early Spaniards had used to head home: the Windward Passage, between Cuba and Haiti; Mona Passage, between the Dominican Republic and Puerto Rico; and Anegada Passage, through the Virgins. Mona and Anegada are too far east for the fuel tanks of the smaller "druggies." Windward is the

most direct route to the East Coast but, at the time, was full of Coast Guard planes and ships trying to dam the flood of Haitian refugees to the United States. So Yucatán was the hot spot, Murtagh said, the best choke point. *Dauntless* would steam down during the day, sweeping the approaches to the passage, then pull back inside the choke point at night, running without lights, a seaborne stakeout.

We turned west, following the keys down the slot between Cuba and the mainland, where the Florida Current runs strong. Murtagh drilled his boat crews, the youngsters who do the boarding and searching. Up on the bridge a thin and serious lieutenant named Larry Yarbrough had the four-to-eight watch. Yarbrough had been aboard two years, longer than anyone else. He arrived a week before the start of the Cuban sealift, the shepherding in spring and early summer of 1980 of the more than one hundred thousand refugees taken from Mariel, on Cuba's northwest coast, to Key West. After World War II, it was the Coast Guard's finest hour, a mayhem of pulling people from swamping boats, tending the sick, keeping the peace, passing out Spanish-language paperback Bibles. "My first thought," Yarbrough said, "was, My God, so many!"

Yarbrough was the ship's operations officer, the man who led the mission. I met him when a patrol boat ferried him to *Dauntless* from Key West, where he had been testifying for the prosecution in a case involving one of the cutter's busts, a small freighter carrying twenty tons of Santa Marta Gold. The captain, Yarbrough thought, would get the customary three years. The crew, four thin and dour Colombians who had subsisted for at least part of the voyage on the rear half of a dog, would probably be deported. It is easier, or was then, to convict a captain, since the law presumes he knows what his cargo is. With the crewmen, you must prove that they knew.

Colombian skippers, I was told, generally received a flat fee, out of which they had to pay for fuel, food, and crew. The average hand might get a couple of hundred dollars for a week aboard a filthy, vermin-infested craft rank with the stink of fifty-pound bales of grass. American smugglers tended to make much more, though not as much as what the pilot of a small plane carrying low-volume, high-value cocaine can pocket today.

Yarbrough and Murtagh were a matched pair. Murtagh had the reputation — one he didn't discourage — of thinking like a druggie. When he could, he overflew the area in a patrol plane. He studied videotapes of busts, and he paid attention to intelligence coming in from the Drug Enforcement Agency, the police in Miami, and his own Coast Guard people. His idea of relaxing was to play Risk or Strategy or other games of stalking.

Yarbrough's job was to snap the trap. He stayed close to the navigators around the chart table and listened to the radio traffic. He moved the pieces of the game around in his mind: *Dauntless* here; this contact on the radar there; that mast picked up by Big Eyes, the twenty-five-power binoculars up on the flying bridge, there. If Murtagh moved in to hail, Yarbrough would handle the "evolution," the taking of the prey. He would talk to the suspect vessel, every word recorded or videotaped for playback in court.

Murtagh? "I've got to watch fatigue," he told me. "Sometimes I go thirty-six or forty-eight hours without sleep. My role is not to conduct operations but to act as an observer. But I tend to get too involved," he said with a grin and a shrug, "because I like it."

"I like it too," Yarbrough said. "I start getting bored just before we bust, thinking of all the hours of paperwork ahead. The crew works the other way: they get high when we're busting and grumpy when we're hunting."

Night on the bridge, the best hunting time. Close and black. Whines and yodels from the radios. An occasional command from the officer with the con: "Five degrees left rudder."

The helmsman: "Five degrees left rudder. Passing two-seven-zero."

The officer: "Very well."

And the lookout: "Bridge, I have a contact bearing zero-four-zero, range sixteen thousand."

"Bridge, aye," from the helmsman.

The big binoculars had picked out running lights eight miles off. Murtagh came up on the bridge, bent over the radar, a votary at his altar. Then up to the Big Eyes, hanging there like a U-boat commander at his periscope. Then back to the bridge to confer with Yarbrough and the executive officer. In a minute the squawk box sounded. "Now," announced the executive officer quietly, "set Law Enforcement Phase One." *Dauntless* was going in for a look.

We could see the contact's running lights clearly. He was well over a hundred feet long, heading north. An ensign, the head of the boarding party, came up to talk with the Old Man. He was blond and good-looking, with wild eyes, wearing dark blue coveralls, a life vest, a motorcycle cop's helmet with the Coast Guard insignia. He carried a nightstick and a .45. Men behind him were more heavily armed. In the gloom I could make out a 12-gauge riot shotgun and an automatic rifle. The covers came off the fifty-caliber machine guns abaft the pilothouse, and ammunition boxes clicked into place. "Remember," Murtagh said to the ensign, "if you board, be polite but firm. We don't want anyone hurt, them or us."

We ran down the contact's port side a half mile away and then turned and came up his wake. (More than one smuggler has been convicted on evidence of bales of marijuana and other jetsam fished from the sea.) We were trailing a thousand yards back and closing. "Now the searchlight," Murtagh said. The beam fired across the sea, flipped past the contact, then settled on a white-hulled freighter caught like a deer on a dark highway. Yarbrough motioned his recording-machine man into position and then called on the radio: "Vessel off my starboard bow, this is the United States Coast Guard, over." Nothing. Someone stepped over to the microphone and said the same thing in Spanish. Nothing. Then, in English, came back: "Good evening, gentlemen. Nice to see you out here." It was an islander's voice, rising and falling, that of Captain Carl Miles, Trinidadian, of the *Unicorn III*. Miles said he was carrying orange concentrate from Belize to Tampa, that *Unicorn* was registered in Panama.

I was told that under international law, the Coast Guard could board a U.S.-flagged vessel anywhere on the high seas, but a vessel claiming foreign registry could not be boarded unless the captain gave permission or unless the country of registry agreed to the boarding. Yarbrough asked permission and Miles agreed. We moved to Law Enforcement Phase Two.

The vessel looked innocent in the white light, but Murtagh noted that the life rings bore no name and that it looked as if a name and home port on the bow had been painted out. The boarding party climbed into the boat, the falls moaned, the boat hit the water, slid around our stern, and headed for the freighter. It took a half hour or so, during which *Dauntless* stood off *Unicorn*'s port quarter and the boarding boat off her starboard, for crossfire purposes. Then the searchers called in. *Unicorn* was clean. "Well," Murtagh said, "it's nice to see people out here doing a legitimate job."

"Set the sea watch," he told Yarbrough and grinned at him as he went below. "The night is still young."

Over the next couple of days, *Dauntless* found nothing aboard a lovely ketch and a dingy shrimper. Morale turned damp. Murtagh and Yarbrough refreshed their spirits by showing me videotapes of past busts. First was a freighter. "Look at all those antennae and the radars," Murtagh said. "Much more than any honest skipper would need. The sides are rubbed up; they've been off-loading in a sea. And the fresh paint up in the bow: we look for that as evidence of a name change."

Next, a yacht whose captain was suckered into running a small load of grass right up to the waiting *Dauntless* while the main freight made an end run. The camera showed two young and attractive couples coming aboard as prisoners. Just before the bust, I was told, one woman had taken a big snort of white powder before she emptied it over the side. She was as high as the masthead when the boarding party got to her.

Last was a large freighter with a false water line to make it look as if he were riding empty. The captain claimed Panamanian registry. "We trailed him for hours," Murtagh said, "until we got word that the Panamanians knew nothing about him. That made him a stateless vessel, subject to our jurisdiction. He started heading into Cuban waters, radioing the port captain at Matanzas that he was being attacked by the U.S. Coast Guard. That was just a maneuver. The Cubans are tough on drugs, and their jails are tough on druggies. But when we came alongside to board, he tried to hit us. Look! There!" The videotape showed the freighter closing fast, and we heard Murtagh's recorded voice: "Back down to avoid ramming." The freighter got away that night but came back out of Cuban waters the next morning, meek as a lamb and clean, except for a few shreds of marijuana clinging to the port side. She had dumped her load. But the captain paid: they got him for endangering government property.

On the fourth day a C-130 Hercules from Miami came down to help. Code talk went back and forth. The plane flew patterns until its fuel got tight, then headed home. "Glad to have given you-all a hand with the fishin'," the pilot radioed.

Murtagh got word from his base that a mother ship might be sneaking up toward the Campeche Bank off Mexico, trying to hide herself among the fishing fleet. Another cutter, *Alert,* was coming down from the north in a pincers play. "We could sure use the chopper now," Murtagh said. *Dauntless* has a big pad astern for a helicopter, which can extend the search path out to forty miles on a side, far beyond effective radar range. But the Coast Guard, then as now, was strapped, hard-pressed to pay for fuel for the aircraft that were still in flying shape. Maintenance, patrols, and other cutter operations had also been cut back. *Dauntless* showed the effects of deferred upkeep. After an hour at flank speed, a seawater line broke on the number two engine, and we limped along while the engineers fixed it. In the end, that night we picked up a bright contact on the radar and ran toward it, only to find *Alert.* Two cutters stalking each other on the black Caribbean.

And then luck changed. Late in the afternoon of the fifth day, the lookout spotted a small craft to the westward, heading north.

Radar was blinded by the rain squalls marching eastward across the channel, so Murtagh spent most of his time at the Big Eyes. "Hot dog," said the lookout. "I've got the whole watch to play with this one."

It didn't take nearly that long. We rounded and came up on the stern of the *Daisy Marivel.* Yarbrough tried radio and bullhorn, but nobody answered. Nobody came out on deck. Then a man broke from the cabin with a heavy garbage bag and dropped it into the sea. Another followed with what Murtagh later guessed were bags of cocaine and Quaaludes. Papers floated by us. We tried to net them, but they sank.

Finally, a young voice — the captain — answered Yarbrough on the radio. He didn't know the nationality of his vessel, he said. He was Canadian. His crew was "of Spanish persuasion." He would not permit boarding. However, he had no choice in the matter: "Miami" was painted on the boat's transom as her home port. She was American.

"*Daisy Marivel!*" Yarbrough shouted. "Stop throwing things overboard!"

"We're not doing anything," said the Canadian and swung his wheel hard, sending his boat right under our bows. *Dauntless* veered off, giving the crew of *Daisy Marivel* a little more time to destroy evidence, but in five minutes the boarding party had them, two Cuban-Americans, two Colombians, and one very young Canadian with an auburn Afro and a sharp nose. Another five minutes and Murtagh heard what he wanted to hear: the boat was stuffed with grass, more than ten tons, worth at least $6 million in the street prices prevailing then. Small but satisfying. The forty-ninth bust.

The boarding party returned with its prisoners as night dropped. The smugglers were quiet, cooperative. Advised of their rights, they were photographed and led away to the fantail for a strip search that included body cavities. Then they were given magazines and dry clothes and blankets, and were secured by leg irons attached to a cable running along the deck, out of the weather. They would henceforth be guarded by crewmen carrying nightsticks. With these precautions, there was little risk of violence or of someone breaking loose and sabotaging the ship.

Alert, returning to Miami, escorted Murtagh's prize and its crew north. *Daisy Marivel* was turned over to customs, which prepared her for forfeiture and eventual sale at auction. I doubt she brought much. Drug Enforcement Agency people met the prisoners and processed them for prosecution. Some bales of grass went to court as evidence; the rest went to Orlando for burning in the government incinerator there.

Dauntless did what cutters have been doing for a couple of centuries. She went back on patrol.

5

Reading the Sea

THE ARABS CALLED THE ATLANTIC the Green Sea of Darkness; the Europeans, Mare Tenebrosum. Before Columbus found his way across it, ignorance and imagination had constructed a marvelous ocean. Europe fed on marvels, wrote the historian J. H. Parry, for it was a time of disease and depression, just right for "splendid mendacity." Prester John, the mythical Christian king, was around (Columbus thought he might have crossed his trail in Cuba), and so were monsters and mythical islands by the score. These last didn't disappear from maritime charts until the nineteenth century.

Ocean circulation was beyond comprehension then. There were stories of holes in the poles, enormous caverns at the base of great polar cliffs where the sea whirled and plunged into the earth, "afterward," as one philosopher argued, "to return again to the heads of the fountains and rivers." Those stories persisted for centuries. The better educated thought that the wind and waters followed the stars in their revolutions around the earth. Sailors knew the winds, at least those in the eastern North Atlantic, and Columbus learned quickly that the most persistent of those winds carried across the full sea. Most coastal folk in northern Europe knew that *something* was carrying oddments to their shores, strange things, tree limbs and nuts and seeds. Even now we are entranced by these foundlings on the strand, escaped from the alien sea.

The coast dwellers of the eastern North Atlantic called some of the nuts eagle stones. Men and women and beasts wore them against infertility or the evil eye. I have one of them, from John Dennis, a student of drift seeds. It is *Entada gigas*, the sea heart. Large, flat, and glossy, the sea heart is the fruit of a vine that grows in the Caribbean basin. It can float in the sea for much longer than the fifteen months or so required to make the trip to northern Europe, where it was used in childbirth, to hasten delivery and calm the pains. The midwife placed it in her patient's hand and walked around her, following the sun, saying, "Behold, Virgin, the woman on the sod of death. Behold her thyself, Son, for thine is the power to release the child and succor the woman."

Not much thought was given then to currents that might have carried the flotsam. Mariners dealt with currents all the time in their coasting. They had the land as reference and could thus judge set and speed fairly accurately. But on blue water they had only their sixth sense to tell them if they were being carried off course. Rips and weedlines were their telltales, as were changes in the way their ships answered the helm. And as they drew up buckets for washing, a change in water temperature could be a sign that they had passed into a different oceanic regime. The story is told that John Cabot learned he was sailing in the northern domain of the blue god when the beer in his hold turned warm.

In the ninth century the Arabs, according to Johann Georg Kohl, my Gulf Stream chronicler, were familiar with the Stream's Pacific cousin, the Kuroshio (Kohl writes it Kara Siroo). But there is no record of man's familiarity with the currents of the Gulf Stream gyre until Columbus set sail. On his first voyage he lowered his sounding line, and the angle told him he was being set down to the south and west. And when he found what appeared to be products of Spain washed up on West Indian beaches, he explained their presence in terms of the currents he had sounded. On his third voyage he reported that he was carried rapidly westward from what he called the Mouths of the Dragon, at the entrance to the Gulf of Paria, between Trinidad and Venezuela. That current we know as the Guyana, part of the South Equatorial Current that is deflected north into the Caribbean by the bulge of Brazil.

In 1502, on the Admiral's fourth and last voyage, he struck the Central American coast at what is now Honduras and headed east. In his history of the New World, a term he evidently coined, Peter Martyr says Columbus "found the course of the water so vehement and furious against the fore part of his ship that he could at no point touch ground with his sounding plummet. . . . He affirmeth also that he could never in one day with a good wynde wynn one mile of the course of the waters" as they approached the Isthmus of Panama before turning north toward the Gulf of Mexico.

Antón de Alaminos must have felt the downstream effects of those waters as he traversed the Gulf of Mexico; he certainly hitched rides on them in later voyages. And we know what he did with the Stream. But little detailed information about currents shows up except in the occasional and usually proprietary sailing instruction. For a couple of centuries thereafter, maritime charts burgeoned with news about recently discovered sandbars and other obstacles to navigation, but the arrows denoting major movements were all but absent well into the eighteenth century. Meanwhile, mapmakers insisted on paying tribute to nonexistent circulation, particularly that mythical whirlpool off Norway, the Maelstrom. About all there was to mark the Gulf Stream were the words, usually in Latin, announcing that the "Bahamian Canal" always flows north.

But mariners, on their own, were getting the hang of the Stream. The sounding lead was replaced by the ship's soup kettle or another heavy container, to the improvement of current measurements there and elsewhere in the Atlantic. Sir Humphrey Gilbert reported that weighting sails and sinking them in currents was a fine way of getting drift data. Explorers came back with tales of finding warm water and strong currents off the Canadian Maritimes, isolated evidences of the gyre at work. Still, until the eighteenth century there were no recorded efforts to collect ships' logs, to interview masters, or to find other ways to present a cohesive picture of water movements.

Theory, of course, had little chance to grow in that soil. Not so supposition. There was, as noted, the holes-in-the-poles idea. As the geography of the Atlantic coasts became more familiar, the discharge of large rivers came to be a popular explanation for the cause of the Stream and other major currents. For a while, there was debate over whether North America wasn't really a series of very large islands through which the Equatorial Current flowed. Peter Martyr thought so. "I do consider," he said, "that there might be certain open places whereby the waters should then continuously pass from the East into the West, which waters I suppose to be driven about the Globe of the earth by the incessant moving and impulses of the Heavens."

In the mid-sixteenth century, the searchers after the Northwest Passage to the Pacific, over the top of Canada, argued that the Gulf Stream turned west above Newfoundland and flowed into the "Southern Sea," the Pacific. Sir Humphrey Gilbert took that view, having rejected an alternate track in which the current "must needs strike over upon the coasts of Iceland, Norway and Finmark." Martin Frobisher, whom Gilbert introduced to the Northwest Passage, made three voyages to and around Baffin Bay but could find nothing that interested the British. He contributed a good deal, though, to the hydrography of the North Atlantic by adding weight to the theory Gilbert had discarded. In 1578 he was sailing to the northwest of Ireland and encountered a "great current from out of the Southwest" which set him off course. He thought the current was flowing toward "Norway and other Northeast parts of the world." But then he took a tack for the worse:

> We may be induced to believe that this is the same which the Portugals meete at Cabo de buona Speranza [Cape of Good Hope], where stryking over from thence to the Straits of Magellan, and finding no passage there for the narrownesse of the sayd Straites, runneth along

> into the Great Bay of Mexico, where also having a let of lande, it is forced to stryke back again towards the northeast.

Hakluyt, reading that, scribbled "Mark this current!" in the margin.

Knowledge of the North Atlantic subtropical gyre was accumulating, however. It had to be, what with the annual sailings of the Spanish treasure fleets and the raidings of Drake and other British adventurers. Kohl says these sea robbers

> were carried by the equatorial currents and the trade winds to the west and usually entered the West Indian empire through the passages of Dominica, Guadelupe, etc. — swept with the circuitous trending of the currents through the Caribbean Gulf, landing, burning and plundering here and there, passed with the currents on the Strait of Yucatan into the Gulf of Mexico from which they, laden with booty, made their escape through the Strait of Florida, disemboquing from it with the Gulf Stream into the regions of winds and currents from the west.

(Kohl, as one can see even in this indifferent translation, was not particularly gifted in his prose style. Charles Dickens remarked that "the indefatigable Mr. Kohl is always instructive [but] sometimes tedious.")

Naturalists paid more attention to the provenance of sea beans. A few brave ones went against the folk wisdom that these beans came from trees growing in the depths and insisted they were the fruit of land plants. Some, thinking they came from the Spice Islands in the far Pacific, took to calling them Molucca beans, a name that still sticks. Finally, in 1696, Hans Sloan, a British botanist, recognized in the wrack on Irish and Scottish beaches seeds he knew by firsthand experience to grow in Jamaica.

Theory took on some new dimensions. Investigators understood by the latter part of the seventeenth century that the sun had a good deal to do with the motions of wind and water. Isaac Vossius, a German investigator, thought that solar radiation heated the equatorial North Atlantic to such an extent that water density plummeted and surface waters expanded. The result, he said, was a long ridge of water which ships had some difficulty ascending as they approached the equator. The sun carried this "mountain of a wave" west, where it was deflected north by the continents. Not so, said the French hydrographer Georges Fournier. The very opposite happens: the sun causes such evaporation of the tropic sea that it creates a deep valley which oceanic circulation tries to fill. Both men, of course, were hyperbolic. But as so often happens in science, they left concepts that showed up in later theories, in this case theories explaining variations in sea-surface height. And Vossius had at least a basic idea of North Atlantic circulation. "A ship without sails and sailors," he wrote, "might be conveyed solely by the force of the currents from the Canary Islands to Brazil and Mexico, coming back from there by way of the Florida Stream toward Europe."

Copernicans, including the German astronomer Johannes Kepler, had already begun applying the force of the earth's rotation to fluid phenomena, saying with some truth that the sea, being loosely coupled to the globe, lagged behind in the race toward the east. In 1678 the first attempt at a global current chart appeared. It didn't identify the Gulf Stream as such (it was primitive; its creator was a holes-in-the-poles man), but it started men thinking of interconnections among oceans and seas. When a French fleet reported that the Florida Current increased its velocity in the face of a stiff wind blowing from the north, a notion developed that the same wind, blowing more out of the northwest, crossed the Gulf of Mexico and forced its waters into the Stream.

The French were also responsible for what appears to be the first estimate of the Florida Current's velocity. In the early 1700s, a French astronomer named Laval judged he had been set north at the rate of one league per hour, a calculation well within the ballpark. It was Laval who bragged about the home route the French had developed. They rode the Stream to the Grand Banks and then struck over, well north of the Azores, in the teeth of westerly weather. They undoubtedly lost many a lunch, but they saved two weeks over travel on the Alaminos Memorial Highway. The route, Laval bragged, "is much more to the taste of our nation, which by nature is impatient and courageous." Well, perhaps, but the French were less dashing in sailing from their possessions on the Mississippi to their possessions in the Caribbean: they sailed all the way up past the Grand Banks and over to the Azores to catch the southern route back to Guadeloupe or Martinique.

When it came to knowing something about the Gulf Stream system and where it went, the sailors stayed well ahead of the scientists in the eighteenth century. (There are even earlier references to the Stream as the Sailors' Current.) A Frenchman named Chabert, sailing home from Canada in 1753, noted currents moving easterly toward the Azores but mistakenly took them for the Labrador Current. The pirates of New Providence, the men who dodged down the Bahamian channels and into the Stream to cut out a fat prize, obviously knew what water moved where. Of somewhat more civilized persuasions were the whalemen, particularly those of Nantucket, who ranged up to the Grand Banks, down to the Bahamas, and out to the Azores.

The establishment paid little heed. British captains reserved as much time for a voyage south down the American coast as they did for a trip across the Atlantic. Coastal surveys undertaken by the admiralty just before the Revolution made but passing mention of the Stream. Mail ships from Falmouth, on the Cornish coast, took a couple of weeks longer than they should have to reach New York. The delay so irked some colonists that they asked their agent in London to investigate. He in turn put the matter to his cousin, a Nantucket whaler, who happened to be in town for a visit. Oh yes, said the whaler, he and his colleagues often chased whales along the edge of the Stream, though they found very few within it. He had seen the mail packets slogging along, stemming the current. He had hailed them a few times, he said, and told them of their plight, but "they were too wise to be counselled by simple American fishermen."

The agent? A Mr. Franklin, from Philadelphia.

*

It is not enough to say that Benjamin Franklin was curious. He elevated curiosity to high art. Carl Van Doren, one of his most noted biographers, tells us that at the time Franklin consulted with his cousin, whose name was Timothy Folger, he was rounding up observations to be published in England and the colonies, where he was about to be elected president of the American Philosophical Society. The subjects of his observations included not only electricity — the key and the kite had already caught the public mind — but also "population, smallpox, whirlwinds and waterspouts, geology, evaporation, salt mines, Scottish tunes and modern music, the origin of north-east storms in America, sound, tides in rivers, insects, the absorbtion of heat by different colors, the effect of oil on water."

No one line of inquiry could ever satisfy such a yearning to know, but the Gulf Stream stood fairly high on the list of Franklin's passions. He made eight transatlantic crossings during his long life, and on several of them was often busily deploying some instrument or other over the side. His instrument of choice was the sea thermometer. The thermometer itself had been invented some time before the middle of the sixteenth century and had been improved by Gabriel Fahrenheit at the beginning of the eighteenth. With a maritime adaptation, Franklin became a marine geographer. In the year of the Declaration of Independence, he sailed for France — and took time to track a temperature signal he thought was the Gulf Stream as far east as the Bay of Biscay, immediately off the French coast. Along the way he noted that he saw more gulfweed inside the Stream than outside. Years later, bound for home from the English Channel, he reported that his vessel had been helped so much by countercurrents south of the main flow that she arrived on the American coast several days ahead of schedule.

Franklin knew of the Gulf Stream well before he met Folger. He was interested in reported — and often greatly exaggerated — differences in sea level (some of which may have been caused by tides). In a letter dated May 27, 1762, he wrote,

> It is well known that strong Winds have a considerable Effect on the Surface of the Seas. The Trade Wind blowing over the Atlantic Ocean constantly from the East, between the Tropics, carries a Current to the American Coast, and raises the water there above its natural level. From thence it flows off, thro' the Gulf of Mexico, and all along the North American Coast to and beyond the Banks of Newfoundland in a strong current, called by Seamen *the Gulph Stream.*

And, Franklin continues confidently,

> In those Northern Latitudes the Winds blowing almost constantly North-west . . . the Waters are mov'd away from the North American Coast towards the Coasts of Spain and Africa, whence they get again into the Power of the Trade Winds, and continue the Circulation. Thus the North West Winds keep the level of the Sea lower in the North East Seas of America, as the Easterly Trade Winds accumulate it on the Coast between the Tropics.

Whether or not one could assign the cause of such differences in sea level, said the philosopher from Philadelphia, "the Fact must nevertheless be allow'd; since so long and so strong a Current as that of the *Gulph Stream,* thro' all the Latitudes of variable Winds, can only be accounted for by its having a considerable Descent, and moving from Parts where the Water is higher, to Parts where it is lower." A straightforward explanation if not a particularly dynamic one.

Franklin didn't merely listen to Folger. He asked his cousin to trace the course of the Stream as he knew it, "from its first coming out of the Gulph, where it is narrowest and strongest; till it turns away to go to the Southward of the Western Islands [the Azores], where it is broader and weaker." He asked him to write instructions on how to steer clear of it on the run from England to the northern colonies. Folger did a fine job, in effect outlining the envelope of the Stream, the limits to its writhings. The courses he prescribed introduced foreign skippers to the art of slipping between the Stream and the shallows that still threaten shipping, from Sable Island to Georges Bank to the Nantucket Shoals.

Three versions of the Franklin-Folger chart appeared, the first in London, printed in 1769 or 1770, the second in Paris in 1778, and the third in the United States in 1786, this last a distinct modification. It is not clear that the British paid much attention. As rumors of revolution flowed in his homeland, Franklin may well have tried to suppress what he had so eagerly presented to His Majesty's mariners. In France for the cause, he evidently made sure that the right people there knew about the blue god — anything to speed the flow of arms and supplies to the rebels.

Copies of the first chart disappeared, and historians began to doubt it had ever been printed. But in 1978 an American oceanographer found two copies in the Bibliothèque Nationale in Paris. The third edition, engraved by one James Poupard, is probably the most famous, though it is less accurate than its predecessors in many respects: the Stream flows too close to Georges Bank, and Bermuda lies too far west.

Franklin had the Poupard edition printed in the *Transactions* of his beloved American Philosophical Society. I would love to question the editor about that layout. The map is inserted in the middle of an article about a partridge with two hearts, and its upper left corner is taken up with an illustration for an unrelated paper having to do with the migratory paths of Atlantic herring. In at least one later version, the two maps became one.

Poupard had a nice touch. At the lower right corner of his chart he placed a rock rising from the sea, bearing the chart's title. To the left of the rock Neptune stands, up to his waist in his medium, discoursing with a gentleman standing on the strand, wearing a tricornered hat. The tableau was proper: Franklin always considered himself a "landman." On his last Atlantic voyage he showed himself sensitive to the distance separating the seaman and the lubber. Sailors, he wrote somewhat grumpily, forget that "most of their instruments were the invention of landmen. At least the first vessel ever made to go on the water was certainly such."

These words are part of what became a wonderful work. Franklin, writing it aboard the London packet, ascribed its length to the garrulity of an old man. "As I may never have another occasion of writing on this subject," he said, "I think I may as well now, once for all, empty my nautical budget." Which he did. The result, published in 1786 under the title "Maritime Observations," appeared, along with the Poupard chart of the Stream, in *Transactions*.

This was Franklin's budget: proposals for rerigging ships with many small sails; for avoiding anchor cable breakage; for adopting the old Chinese practice of dividing holds into watertight compartments; for a range of other devices and practices for improving safety at sea. ("Our seafaring people are brave," he wrote in a comment that every Rambo today might well heed. "They reject such [safety] precautions, being cowards only in one sense, that of *fearing* to be *thought afraid*.") He advocated lightning rods aboard ships, catamaran hulls, sea anchors, oven-drying vegetables to preserve them during long voyages. He invented a soup bowl for foul weather, one that would not react to a roll by spilling its contents over the nearest lap. My favorite entry concerns a kite, surely one of Franklin's favorite devices. This one was to be fashioned by a man who had fallen overboard, out of his handkerchief and a pair of crossed sticks — their provenance unknown. Held just right, said Franklin, the device would aid the swimmer in a "long traverse." I see it as a great bandana raised to the gale, towing its owner — who is planing along on his backside — over the horizon.

The Gulf Stream received full treatment in "Maritime Observations." Franklin retold the Folger story, forgetting not one detail. He described, with appropriate tables, his work with the sea thermometer and explained his theories establishing wind stress as the principal cause of the current. He turned to weather over the Stream. The ascent of warm air drew in cooler air from the sides, he said. The collision of such inrushes formed "those tornados and waterspouts frequently met with" on the current. Off Newfoundland, that same warm air condensed into pea soupers, "for which those parts are so remarkable."

Others around Franklin's time tried their hands at putting the

Stream on paper. Walter Hoxton, a captain in the Maryland tobacco trade, did so in 1735 after carefully noting how the current set him north as he crossed it on his runs to and from England. In 1772 a surveyor in the southern colonies, William DeBrahm, followed the "velocious Stream of Torride Mexico and Florida" to a point somewhere off Delaware and then extrapolated from there; he had his subject overrunning Georges Bank and generally pressing so close to the coast that the chart might well have scared ships from England into turning south too soon, right into the teeth of the real current. Nonetheless, the "loss" of Franklin's first chart gave DeBrahm a short moment in the cartographic sun.

In the early nineteenth century, ideas of a wind-driven Stream similar to those Franklin had espoused were given new life by British geographer James Rennell, whose major work was published in 1832, shortly after his death. Rennell's thinking influenced the field of Gulf Stream study for years. He divided oceanic movement into drift currents, caused directly by wind action, and stream currents, caused when drift currents ran into a sizable obstacle, like the North American continent, and were diverted. The Stream was a stream, he said. He subscribed, as Franklin and so many others had, to the erroneous notion that the Stream's engine was a hydraulic head in the Gulf of Mexico. But then Rennell parted company with Franklin. The Stream did curve east toward Europe, thanks in part to the earth's rotation, but except for occasional intrusions, it didn't get there. What did get there was mostly wind drift. The Stream itself wandered into the Sargasso Sea and disappeared. Further, he said, warm water did not always translate into Stream water; it could be an eddy or an overflowing on the surface. So much for thermometrical navigation.

The chronometer, key not only to good navigation but to good oceanography, wasn't in wide use until after Rennell's death. With it, scientists could check longitude against dead reckoning and come up with a better sense of east-west currents. And if the thermometer was no longer thought of as a positioning instrument, its use in obtaining thermal profiles of the sea became a central part of marine science. It still is.

Using these devices, adventurous souls began to study circulation in the far northern Atlantic. The explorer Alexander von Humboldt, he of the Humboldt Current off the west coast of South America, advanced the idea that the seas between Norway and Spitzbergen were "warmed by a current from the South-West," to the extent that "navigation is uninterrupted even in the midst of the strongest winter." A Danish captain named Irminger, whose name survives in a current that loops up south of Iceland and around the southern tip of Greenland, found anomalously warm temperatures in that region. Exploring in the Faroes, between Iceland and Scotland, he came across islanders who showed him part of an American canoe and told him they used to get a lot of driftwood from America. They burned it and built houses with it and were convinced it came down the Mississippi into the Gulf of Mexico and on over to them. The supply was dwindling, they said sadly, probably because settlers in the Mississippi valley were cutting down so many trees.

James Rennell worked extensively with records of Atlantic navigation. So did American naval lieutenant Matthew Maury, of "there is a river in the ocean" fame. Maury made use of ships' papers at the navy's Depot of Charts and Instruments to develop his physical geography of the sea. Naval reports came to him as a matter of course; those from privately owned vessels, still guarded almost as closely as the Spanish had guarded their rutters, were forthcoming only after Maury promised to send owners free copies of the charts drawn from their information.

Unlike Rennell and Franklin, Maury discounted wind action as the dominant cause of currents. Surveys in Florida had indicated that sea level on its Gulf coast was only inches higher than that along the Straits, surely not enough to provide the head necessary to power the Stream. No, said Maury, the real source was to be found in density differences caused by unequal insolation and by precipitation and evaporation. Wind helps, he said, but winds can drop. Whereas "the changes of temperature and of saltness and the work of other agents which affect specific gravity of sea water and derange its equilibrium are as ceaseless in their operations as the Sun in his course, and in their effects they are endless." The public read and believed, and oceanography's long debate over the relative dominance of the sea's forcing agents picked up some speed.

Maury could have paid better attention to the advances science was making. He misread Coriolis's arguments about motion on a rotating sphere. He thought, for example, that the Stream sloped toward its sides like a barn roof. Scientists envied him his ability to gain support for his work, but they took a dim view of the work itself — none more so than a gentleman named Alexander Dallas Bache, superintendent of the United States Coast Survey, on its way to becoming the most scientifically oriented agency in the federal government.

Bache had another claim to fame. He was the great-grandson of that peripatetic thermometrician from Philadelphia. The Stream was in the blood of the family Franklin. Ben had taken his son William, and William's son William Temple, to sea to help with his research. Ben's grandnephew, Jonathan Williams, became so enamored of the sea thermometer that he later claimed it could locate not only currents but also "banks, coasts, islands of ice and rocks under water." With that much momentum in the lineage, it was only natural that Alexander Bache should follow along.

The Coast Survey, forefather of the Coast and Geodetic Survey and today's Geological Survey, was supposed to tilt toward the applied and the practical in its research. But Bache, a professor of natural philosophy and the first president of the National Academy of Sciences, bulled ahead on basic research. He convinced Congress of the value to commerce of understanding the Stream, "the great sea mark of the coast." His vessels started running transects perpendicular to the current's axis, returning inshore to check their position. His sea thermometers were worked hard, their sampling increased in areas where temperature variability was greatest. He loaded his scientists with questions: What are the limits of the Stream? What makes the currents shift, and how do you recognize the shifting — by thermometer at the surface or below, by sounding line, by checking animal life, the weather, the salinity of the sea? What are the boundary conditions? How can you determine interchanges of water and other of the system's sleights of hand?

The transects and surveys led Bache to a picture of the Stream as a streaky beast, "banks" of warm and cooler water running parallel to the axis, which was warmest of all. He attributed these variations, mistakenly, to bottom topography, to rises and ridges that interrupted the flow as a stone shoulders the water in the Deerfield River. The streaks could easily have been errors of measurement and navigation, since the instruments Bache's scientists were using were terribly primitive. Drift of ships or floats could not be calculated precisely. Sea thermometers, of British design, registered approximate maximum and minimum temperatures without reference to depth, and tended to crush under the pressure of the sea below five hundred fathoms or so. Water catchers, insulated containers designed to bring water from various depths to the surface for temperature reading, didn't work very well either. Yet the survey persisted — measuring, extrapolating, guessing. Bache talked about the existence of "a counter current of cold water from the pole below the warm current from the equator" a hundred years before it was found, and — a true Franklin — he worked up ideas about how to send sea anchors or floats down to measure it.

Actually, the ephemeral bands of different temperature were reported to Bache by his brother, Lieutenant George Bache. In *The Edge of an Unfamiliar World*, my favorite history of oceanography, Susan Schlee says that George also found phenomena of more lasting value. He apparently initiated the collection and preservation of sea-floor sediment samples, and he described in some detail the sharp western edge of the Stream, which he called "the cold wall." The lieutenant thought the wall was fixed (it actually moves as the Stream shifts position), and he might have corrected his error had he lived. But in September of 1848 a hurricane hit his brig, the *Washington*, off Hatteras as he was returning from a particularly successful survey. The pilot aboard reported he "saw Captain Bache and several men overboard astern, apparently as if stunned or hurt." He tried to get a line to them, he said, but a great sea broke over him, and when he cleared his eyes, the men were gone. *Washington* made it to port, barely.

Maury and Alexander Bache were anomalies in the reports-and-routine of government work, and when they moved on, the oceanographic thrust of their organizations declined. But in 1885 another anomaly came along. Lieutenant John Elliot Pillsbury was a mariner with an engineer's imagination, a tinkerer of Franklinesque proportions, at least when it came to devices for reading the sea. He developed a current meter that worked well, and a method that would deploy it to best advantage. During the mid-1880s he sailed out to the Stream in the steamer *Blake* and anchored in the current. He was not the first to do so, but he was the first to do it repeatedly and to good scientific effect. He wrote a long report about his work, one that in its historical review of Gulf Stream research borrowed extensively from the indefatigable Johann Georg Kohl.

Like Maury, Pillsbury knew how to devise a line that would get public attention. "In a vessel floating on the Gulf Stream," he wrote,

> one sees nothing of the current and knows nothing but what experience tells him; but to be anchored in its depths far out of the sight of land, and to see the mighty torrent rushing past at a speed of miles per hour, day after day and day after day, one begins to think that all the wonders of the earth combined can not equal this one river in the ocean.

At school near the Deerfield River, we used to tie a piece of barn door to a tree overhanging a stretch of good flow and ride it as it bucked the current. I think of that, looking at the illustrations in Pillsbury's report. He and his people used kilometers of wire — they worked in depths of up to 2,180 fathoms, or more than 4,000 meters — buffered against shock by special booms, brakes, and accumulators made out of great stacks of rubber doughnuts. He measured and measured, upstream and down and in the Caribbean.

A true believer in the powers of science, Pillsbury was not daunted by his subject's penchant for acting like an unmanned fire hose. Things might appear erratic, he said, but "probably every motion . . . is absolutely governed by laws which can be ascertained." And, in a startling flash of foresight:

> The moisture and varying temperature of the land depends largely upon the positions of these currents in the ocean, and it is thought that when we know the laws of the latter we will, with the aid of meteorology, be able to say to farmers hundreds of miles distant from the sea, "you will have an abnormal amount of rain during next summer," or, "the winter will be cold and clear," and by these predictions they can plant a crop to suit the circumstances or provide an unusual amount of food for their stock.

Pillsbury's findings provided ammunition for both armies in the winds-versus-density arguments over current causation. He himself tended to be a winds man. But his report came out in 1890, and by that time the United States had shifted its attention away from the sea and toward the development of its great hinterland. Increasingly, the work of oceanography came to be taken up by Europeans.

Much of that work dealt more with organisms and oozes than with the puzzlements of circulation. Charles Darwin became interested in sea beans as exemplars of species dissemination. He collected them and sent them off for identification and germination. Wyville Thomson and others aboard the British research vessel *Challenger* in her world cruise of the 1870s brought home hundreds of specimens of sea life and sediment. Like Pillsbury, Thomson believed that the Gulf Stream, which *Challenger* probed, was "the one natural phenomenon on the surface of the earth whose origin and principal cause, the drift of the trade winds, can be most clearly and easily traced."

By the 1880s, study of the Stream and the arguments it generated had attracted royal attention. Prince Albert I of Monaco loaded his schooner with a cargo of bottles, barrels, and copper globes, each with a message asking the finder to report location and date of find, and sailed off into the Atlantic. He had ballasted his globes and barrels so they would float almost entirely submerged, out of the wind's grasp, and he set them in the sea by the hundreds — off Newfoundland, off the Azores, in mid-ocean. He got back an eighth of what he put out, but the patterns of recovery did indicate the presence of a subtropical gyre and its child, the North Atlantic Current, wandering off toward high latitudes.

Albert went on to do what many Europeans were doing, but with longer-lived success. He built a center for marine research, the Musée Océanographique de Monaco, which is still functioning. The prince himself developed into a Renaissance man of the oceans, sampling, inventing, funding expeditions. It was he who demonstrated that pelagic animals could be caught thousands of feet below the surface, that the sperm whale hunted and caught enormous squid in those depths. And, after World War I, he predicted that German mines torn loose from European harbors would follow the general pattern of his drifters, floating south, then west to America, and finally up the Gulf Stream. Susan Schlee tells us that several dozen of the mines were found to have done just that.

It is only a small distortion of truth to say that fisheries problems have led to the establishment of physical oceanography as we know it. Albert of Monaco was driven in part by those problems, and so were scientists in Scandinavia. But where Albert and others (including the U.S. Coast Survey) concentrated more on attempts to measure circulation directly, with drifters and other instruments, the Scandinavians opted more for water-mass analysis. Because food fish seemed to prefer waters of a particular temperature and salinity range at a certain depth, an understanding of the movement of those waters might be able to ease the lean years, when the shoals of herring no longer fed on familiar bottoms.

At first the Scandinavians worked mostly inshore, in the fjords and along the coast. Soon, however, they were moving away to deep water, chasing the mother currents that fed their home circulation. One who went far was a Norwegian explorer, Fridtjof Nansen. In 1893 he sailed north in his specially strengthened vessel *Fram* with the idea of letting the ice take her. The currents would move the ice, and he would study the currents. Three years later Nansen and *Fram* were back, separately; the two had parted company when Nansen tried to sled to the pole, gave up, and did some hard traveling before being picked up by English explorers. To help him calculate the motions *Fram* had encountered, Nansen turned to a Swede, Vagn Walfrid Ekman, who drew from the data a theory that has been used to help explain oceanic circulation ever since.

Nansen told Ekman of long days aboard his small schooner watching the ice drift. It moved to the right of the wind, he said. Ekman noted that and other information, and quickly developed the idea that wind stress, the drag of the air across the ocean surface, produces a current in the water that moves in a direction forty-five degrees to the right of the wind. With depth, the water spirals farther to the right, losing velocity. The mechanism operates through a frictive force whereby each water "layer" (a term of scientific convenience rather than an accurate description) influences the movement of the layer below it.

The Ekman layer, the hundred meters or so of ocean occupied by the spiral, is a useful tool for explaining how the ocean builds insignificant mounds which in turn help build significant currents. In a circulation such as the North Atlantic subtropical gyre, where winds blow toward the west in the southern part (the trades) and toward the east in the northern (the westerlies), Ekman drift drives water toward the interior of the gyre — to the right of the wind.

At the turn of the century another Norwegian, Vilhelm

Bjerknes, gave oceanographers a method of linking their measurements of water properties to measurements of water movement. Like many other early contributors to sea science, Bjerknes was a meteorologist. His fluid behaved in sufficiently similar fashion to seawater to enable him to devise a dynamic theory of oceanic circulation. Bjerknes started with the old saw about water seeking its own level. That level, he said, is determined by a number of forces, the first of which is surface topography (astounding to those who, like me, were brought up thinking that the term *sea level* means what it says). Ekman's wind machine is responsible for some of the sloping. So are massive ridges and trenches in the sea bottom and local variations in gravitational attraction. Bjerknes and his followers said that since the pressure applied to water under such slight convexities was greater than it was in the surrounding sea, water would flow out to areas of lower pressure. As it did so, the Coriolis force would appear, balancing the gravitational flow in such a way that the water curved to the right in the Northern Hemisphere, producing, under perfect conditions, a geostrophic current, and in more normal, hectic conditions, a quasigeostrophic one. The Stream, with its tiny surface slope along Florida, is quasigeostrophic.

Nobody could measure oceanic surface topography then; satellite technology is only now getting the hang of it. To find pressure differences in the sea — the important ones oceanographers call horizontal pressure gradients — the thing to do was to conceive of an area deep down in which water lay still, a level of no motion. Then you went up through the layers, calculating the density of each. Density being a cousin of pressure, if you toted up your sums for, say, two columns of water above the level of no motion, you got a pretty good idea of how the layers in the columns matched up against each other and which column stood above the other at the sea surface. You could then draw something like a geologist's schematic rendering of rock strata and look at the dip and strike of each layer. From such numbers and doodles, the early physical oceanographer could calculate his flows.

There were problems with the dynamic method, as with any other attempt to figure out what goes where in the millions of cubic miles that make up the seventy percent solution. The level of no motion was and is hard to establish. But at sea, science must work most watches in approximations of reality, and Bjerknes's approximation was a welcome one.

With dynamics better in hand, oceanographers turned to the instruments available for measuring density. The sea thermometer had improved greatly since Benjamin Franklin's time, but it needed further refinement if the precise readings required by geostrophy (to hundredths of a degree) were to be made. Water catching, the business of trapping samples at precisely measured depths and retrieving them for analysis of salinity and other chemical properties, was not up to the job either. More sophisticated devices gradually came on the market. One, a reversing water bottle with special thermometers attached, became a workhorse found today aboard almost every research vessel. The first instrument I was permitted to handle on my first research cruise out of Woods Hole, it was named, as usual, for the man who helped make it: the Nansen bottle. Using a hydrocast, which consists of a number of those bottles (or their more modern offspring) clamped to a length of wire, you can get readings of depth, temperature, and salinity good enough to perform dynamical analysis.

European marine science blossomed. Sweden, Norway, Denmark, Ireland, Great Britain, the Netherlands, and Russia joined in 1902 to set up the International Council for the Exploration of the Sea, which in the flow of the years has grown large in membership and missions. In 1925 the Germans took the first full measure of the Atlantic, running hydrocast sections, most of them south of the equator, with their research vessel *Meteor*. Although fisheries management was still a major client, marine science for its own sake had become an acceptable proposition.

The United States lagged somewhat in ocean theory as the century turned. Its great contribution occurred during World War I, when it developed an underwater sonic instrument for detecting submarines, icebergs, and other dangers to Allied shipping. The idea's genesis was the notion of banging bells underwater, a notion that grew into oscillators and receivers. But after the armistice, American marine research went right back into its shell. Oh, there were individuals putting to sea to examine this and that — Henry Bryant Bigelow, for instance. In 1908 Bigelow had begun what turned out to be decades of work studying the dynamics and denizens of the Gulf of Maine. He gravitated there after his mentor, Harvard's famed zoologist Alexander Agassiz, suggested he sail off the New England coast and collect some specimens from the Gulf Stream and the waters inshore. Although he published three landmark monographs on his work, Bigelow had a terrible time aboard ship; as an older man he applied a quaint title to an introductory lecture he gave oceanography students: "Seas I Have Vomited In."

Bigelow saw that not much science was going to get done by individuals working off the odd fisheries vessel or some rich friend's yacht; the ocean had to be institutionalized. Which explains in part why he found himself writing a report for the National Academy of Sciences on oceanography, or the lack of it, in the United States. He could find only a couple of hundred people in the country who qualified as blue-water researchers. No advanced degrees were being offered in the science. There was only one oceanographic institution — Scripps, in Southern California — worthy of the name, and it was still largely biological in its operations. (Today, a well-rounded oceanographic center deals in physical, biological, and chemical oceanography, marine geology and geophysics, ocean engineering, and a number of interdisciplinary projects.) Uncle Sam, who once took more to a sou'wester than to that striped top hat, was clearly growing lubberly.

By design, Bigelow's report went on to the Rockefeller Foundation, whose family patriarch, John D., knew what help oceanography could be in the young business of looking for oil offshore. That pragmatism may have increased the foundation's receptivity to Bigelow's recommendations for substantial aid, including the building of an oceanographic center on the East Coast. The result was a gift of three million dollars, large in those days, even larger given the crash of the stock market shortly before it was made. The new place would have as its first director one H. B. Bigelow, and it would operate only in summer: after all, no gentleman scientist would agree to spend a whole *year* in the little Cape Cod village of Woods Hole, Massachusetts. There are five science centers in Woods Hole now, two of which — the National Marine Fisheries Service facilities and the Marine Biological Laboratory — predate the Woods Hole Oceanographic Institution.

On Water Street in summer, you wouldn't know you were surrounded by what is arguably the world's most outstanding concentration of marine science. Out in the harbors, the white blades of sloops and ketches pierce the wind. Eel Pond, the heart of the village, is choked with hulls of every persuasion. People by the hundreds mosey down the sidewalks, posing for pictures on the drawbridge, looking for the local aquarium, killing time until their ferry leaves for Martha's Vineyard or Nantucket. The brick laboratory buildings are simply backdrop.

The "Oceanographic" — "WHOI" to the informal — started up in the early thirties with one brick building and one ship. The ship was built in Denmark exclusively for blue-water research, and for the first fifteen years of her working life she was the only craft in the United States so designed. She was a sailing vessel with an auxiliary engine, for sail at the time provided a steadier platform and a longer reach between fueling stations. According to historian Susan Schlee, she was the largest steel-hulled ketch of her day. She carried the name *Atlantis*, but in Woods Hole she was the "A-boat," and she was sold to the Argentine navy in the mid-sixties. I never saw her, but I heard plenty about her. About the enormous range of science done on her decks. About the pranks, like the radio report sent back to Woods Hole: "Sea state six; sobriety zero." And about the landlubbers learning ever so slowly to be seamen: "Land ho!" "Where away?" "Far, far away."

The captain of the ketch had the right forename for the job:

Columbus. Iselin was his surname, but the elders at the Oceanographic who remember him aboard *Atlantis* or as director Bigelow's successor call him Columbus and assume the listener knows who's what. The Iselins were well-to-do, and young Columbus had spent a good deal of time at sea, yachting and adventuring. He was seasoned enough to tackle the study of North Atlantic circulation straight out. In little more than four years of research cruising, he produced a paper on the movement of waters in the "Bermuda triangle," which he bounded by lines running from Nova Scotia to Bermuda and then over to the mouth of the Chesapeake. Iselin showed a seaman's humility in addressing his subject. "The problem of oceanic circulation," he wrote, "is such that we cannot hope for a satisfactory solution for a long time to come." But his eye and his hunches produced many conclusions that have survived the remodeling by successive investigations to which all science is heir. He went after the popular concept of the Stream as a hot current in a chilly sea. "The fact is," he wrote, "that by far the greatest part of the [Stream's] water is relatively cold, while even the southeast side of the current (except at the surface) is not quite as warm as the water at corresponding depths in the Sargasso."

Iselin coined the names oceanographers now use for the parts of what he called the Gulf Stream system: the Florida Current, from Tortugas to Hatteras; the Gulf Stream up to the Grand Banks; and then a number of bifurcations as associated currents moved generally eastward. It would be difficult, he thought, to study those transatlantic flows, to identify the courses of those currents, because they are hidden beneath a shallow sheet of water moving under the wind known as the North Atlantic Drift.

Iselin was also struck by the similarity between the temperature-salinity relationships of water measured vertically and water found on the surface as he sailed *Atlantis* north to south in late winter. The vertical measurements that interested him were those taken through the main thermocline, that region of the ocean usually found from a couple of hundred to more than a thousand meters down, in which temperatures drop most rapidly with depth. Above the thermocline, waters are usually relatively warm and well mixed by the wind. Below it the sea is colder and more stratified. What that meant was that water masses are "formed" on the surface in the cold months. They then carry the signatures of their formation — specific temperatures, salinities, trace gases, and other properties — with them as they sink and flow to their proper levels in the North Atlantic subtropical gyre.

The "A-boat" carried scientists across a million and a half miles of ocean in her third of a century at Woods Hole. I say scientists, but many aboard were laymen, there because of the Woods Hole version of the press gang — the pleasant-sounding fellow who, after assuring himself of the strength of your back, says, "Say, you've got some free time. Would you like to go to sea?" No pay, you understand, or very little, but free food and lodging and a lovely view. The Oceanographic was too poor to offer more.

One who answered this question in the affirmative was Frederick Fuglister, Fritz to his colleagues at the Oceanographic. Fuglister was from Washington, D.C., a musician and a painter. The Mexican muralists interested him, and he was developing that art, painting panels on the walls of police stations and post offices for the WPA, when one of Columbus Iselin's lieutenants made his pitch. Off went the muralist to the fisheries of Georges Bank, where he threw up and fell in love with the sea.

A week here, a month there, and Iselin offered Fuglister the job of tending to a machine, the bathythermograph, developed in the late thirties. The device functioned by dropping through a couple of hundred meters of sea while a small stylus scratched a profile of temperature with depth on a slide of smoked glass. The BT gave dynamic measurements a great lift: it took far less time than did the hydrocast, with its clamping and unclamping of bottles, and unlike the hydrocast, it could be used when a ship was under way.

Fuglister became a master of the BT and what it produced. The United States was at war by then, and BT slides from all over the Allied Atlantic came flooding in for him to photograph and analyze. In time he knew more about the surface layer temperature than just about anybody else. But Fuglister's parents were Swiss, and, besides, he had been an *artist*. That seemed enough, as he tells it, for the navy to delay giving him appropriate security clearance until the war was well along. Fuglister used his artistic training to prepare temperature charts that, since variations in temperature can change the velocity and direction of sound under water, were invaluable to those who ran Allied submarines or hunted Axis ones. Once the charts were finished, they were whisked away to the military. Their creator didn't have the clearance to look at his own work.

Fuglister took that nonsense, and the arrogance those with advanced degrees so often display toward those without (he was once introduced by the head of another oceanographic institution as a prime example of "Iselin's idea that anybody can be an oceanographer"). When the war ended, he decided he was going to tackle the biggest thermal anomaly in his ocean. He had the BT as much under control as a man can ever control things at sea, and he had Loran, the long-range navigational system developed during the war. The latter, he thought, would tell him where he was while the BT told him where the Gulf Stream was.

"What I wanted to do, all along, was to follow the Gulf Stream right on over to England, where it warms the beaches," Fritz boomed when I asked him about his pursuit of the blue god. "I never did that." He laughed, his mouth screened by a great frizzle of beard. "I spent years and years trying to get beyond the Grand Banks, trying to follow a strong current and make sure it was always the same." Downstream from Hatteras, Fuglister found great bends in the Stream — eddies, what have you. He found that the Stream itself could shift position by several miles in a day. He was learning what oceanographers today sadly accept: that the Gulf Stream is so strong it can make its own rules. He came back from each cruise, on *Atlantis* or on other vessels of the Oceanographic's growing research fleet, with piles of data that told him that maximum current velocities generally occurred in summer upstream and in winter downstream and that minima occurred everywhere in the fall; that on leaving Hatteras the flow of the gyre, like a transatlantic ship, follows a great circle route toward (remember, not *to*) Europe, while the American coast falls away to the north. He and a colleague, Bruce Warren, reported that large meanders were apt to develop in the patch of water where an arc of volcanoes, the New England Seamounts, rose to within a thousand or two thousand meters of the surface.

Multiship cruises came along, giving oceanographers a wider view of what was going on at a given time. Fuglister helped plan and went on several, including one he referred to as the DREAM cruise. The navy, which was funding a good deal of the Oceanographic's work (by then year-round), changed that to Operation Cabot, a pretty good name for research in an area near the Grand Banks. But Fuglister was saddened by the shift. His acronym had stood for Damnedest Results Ever Achieved by Man.

Fuglister wrote his fair share of papers, but he told me that he never caught the scientist's fever to publish. Perhaps the painter in him rebelled at the thought of so much paper being covered by mere words. "It's all theory now," he said to me. He chuckled, and his beard vibrated. "Simplified, simplified, simplified."

For years Fuglister worked with a colleague named Valentine Worthington, a man of cultivated accent who neglected his classics studies at Princeton well enough and long enough to flunk out. He could have returned, but the war came, and afterward he thought he'd try for a job at the Oceanographic. Columbus Iselin assigned him to some projects Fuglister was operating. At one point, Worthington went up to St. John's, Newfoundland, to board *Atlantis* and look after things on the next leg of a research cruise. Fuglister told him to follow the Gulf Stream to Europe. "But as we all know," he told me recently, "the Gulf Stream doesn't go to Europe, and I had certain difficulty."

That is Worthington's wry way of scoring a point for his side, his controversial argument developed during months and months at sea. The argument holds that most of the water entrained in the Stream, shallow and deep, recirculates in a tight gyre in the northwest quadrant of the Atlantic. To the north of that is yet another gyre in the depths, half as energetic but also rotating anticyclonically (clockwise, in the Northern Hemisphere).

Worthington worked with water masses and their differing properties. When he found that there was a lot more oxygen in the North Atlantic Current off the Grand Banks than in the Gulf Stream proper, he concluded that the two had to have different sources and that there were, in fact, two gyres. To him, the North Atlantic is a fjord in its basin contours, dammed to the north by high sills that separate it from the polar sea. You must account for water movements within that fjord in such a way that masses with a certain temperature and chemical content don't invade masses with different ones. Worthington did that to his satisfaction. Looking back on the controversy over his conclusions, he wrote that "the main objection to my circulation scheme for the Atlantic is that it is not in geostrophic balance. I was not able to conserve mass geostrophically without grossly violating water mass boundaries or moving water through solid pieces of continent."

James Rennell, of course, would have agreed with Worthington's refusal to extend the Gulf Stream eastward to warm Europe. So would many of Rennell's successors down to the present. The salinities and other properties of Stream water don't match well with those of British waters, goes one argument. Baleen whales, not found in the Stream, have been plentiful around Britain. British meteorologists long ago established that local sea-surface temperatures were linked to variations in the temperature of the air over them. That, said one scientist, undoubtedly glaring through his monocle, "is quite fatal to the theory of the British Seas being dependent on an influx of warm water from the Gulf Stream, for had that taken place, any trifling variations in the temperature of the air would have failed to affect that of the sea."

When I last saw him a few years ago, Worthington, looking

like an aging British actor, was wearing a T-shirt with Japanese ideograms on it. "That's 'Cold Wind — Two Gyres,' " he said. "It pays to advertise." The cold wind? Worthington believes that outbreaks of icy polar air are what stimulate the currents of the western North Atlantic. The outbreaks chill the surface of the Sargasso Sea, which slides off to the south. In the endless balancings of the earth's heat budget, that triggers an increase in the flow of warm water north, through the Gulf Stream.

Worthington is still convinced he's correct in the heart of his arguments. Still, he says, everybody's ocean is different, and if you don't watch out you can let pride in your work trip you up: "I don't think it matters if I'm right or not, really. I think the point of the matter is always, Are the observations any good?"

Like Fuglister and Worthington, Henry Stommel has no advanced degree. It is typical of him that when some of his peers started calling him the finest physical oceanographer of our time, he thought seriously of going back to school and getting one. Stommel came to Woods Hole in 1944. He had graduated from Yale four years before and seemed headed for the ministry, but decided to change course. "Perhaps he realized," a friend suggested decades later, "that there could be little opportunity to develop simple models in that area, and even less likelihood that he could test his ideas against data." Stommel did some graduate work in astronomy but eventually applied to the Oceanographic. When he went to the Cape he did so as a conscientious objector, no small step amid the patriotism and propaganda of World War II.

Stommel was a natural for physical oceanography. He was not particularly strong in advanced mathematics, but that mattered less then than it does now. He did have an innate sense of the physics of things and an uncanny ability to hunt down sound explanations of physical processes. And that was what interested him most, the processes of the ocean. Descriptive oceanography, the order of the day when he first arrived at the Oceanographic and still emphasized now, was all very well — you had to know what went where — but to Stommel the *how* of motion was the moth's flame.

There is something of Pan in Stommel's face, in his behavior. At one time or another he has painted, written, cooked, experimented with explosives, farmed, collected music boxes, examined the hydrodynamics of the plumbing in his lovely old farmhouse inland from Woods Hole, pulled many an inspired prank on his colleagues, all the while building constructs of the sea in his mind. He has built a model railroad on his place for his grandchildren and keeps it in good shape, and when I visited him a few years ago, he told me he was working on a children's story called *Cousin Tom's Field Book on Survival at Night for Cats.* He was going to make it a little bit grisly, he said, because kids like that.

Stommel wasn't long at the Oceanographic before he began to do some serious thinking about the blue god. At one point he was having difficulties with a project that called for deployment of drifting floats. The floats wouldn't stay in the current, and the Oceanographic's business office was after him about the project's expenses. "When I got that hassle," Stommel told me, "I said, 'Hell's bells! I'm going to give up trying to make these observations. I'm going to write a book about the Gulf Stream.' "

Stommel had already written a paper called "The Westward Intensification of Wind-driven Ocean Currents," a seminal piece of work in ocean dynamics. In it, he conjured up a simple North Atlantic, its curvatures flattened out in a rectangle, and he demonstrated something that giants of the day had failed to recognize: If you think of the Coriolis effect, the curvings of fluid in motion on a rotating earth, as constant with latitude, the simple model of the ocean gyrates as the layman would expect, with the center of the gyre in the center of the model. But if you allow the Coriolis effect to vary, to increase with latitude, the spin of water parcels in the sea increases also, the so-called Beta effect, and that produces a shifting west, a crunching of flow lines against the western boundary — a Gulf Stream. That one paper, of the hundred or more Stommel has published, was enough to give him recognition throughout the small global fraternity of sea scientists.

The Gulf Stream was not Stommel's first book. In 1945 he had published a popular work called *Science of the Seven Seas*. That earned him enough money so he could sail off to Europe to collaborate with scientist friends in studies such as investigating eddies by tossing pieces of parsnip into a lake. He also had a pamphlet to his credit, a piece of work turned out on his printing press at home and entitled "Why Do Our Ideas about the Ocean Have Such a Peculiarly Dream-like Quality?" But *The Gulf Stream*, published in 1958, was a first attempt at collection and synthesis of ideas, the author's and others', about a central problem in oceanography. Stommel reviewed the history of Gulf Stream research from Franklin on up, and he presented his own theories, including one that carried considerable drama.

That theory dealt with abyssal circulation, the movement of water on or near the bottom of ocean basins. Oceanographers knew that warm water is advected northward from the tropics, but no one had really demonstrated much of what happened after that. What goes north must come south to conserve mass — or, as Val Worthington says, to keep Europe from drowning. There are zones at high latitudes where chilled water sinks, especially in winter. How does that convective flow return south? Working with the creator of an elegant rotating water tank, Stommel concluded that the answer was a relatively narrow, relatively fast deep current that flowed adjacent to but counter to the Gulf Stream. The Stream itself flowed through roughly the upper half of the water column while the undercurrent, hugging the continental slope of North America, ran south, crossing under the Stream as it flowed northeastward off the Blake Plateau into open ocean, and continuing down into the South Atlantic.

The business of testing the hypothesis began. Looking back on the process, Stommel wrote, "You rack your brains for a testable idea, and then you get so involved in it that you begin to hope that it won't be knocked down — even if you are about to learn something else." The deep undercurrent wasn't knocked down. Just before *The Gulf Stream* was to be published, Stommel received wonderful word: Worthington and another old friend, John Swallow of Britain, inventor of a tubular metal device that could be made to float with the current at desired depths, had found evidence that Stommel was right. They had deployed some of Swallow's floats off the coast of South Carolina, and most of those set at 2,500 to 2,800 meters had trundled away in a southerly direction at speeds of a tenth to a third of a knot — practically jet propulsion in the world of abyssal circulation as it was then perceived. And Worthington's hydrocasts in the area found that the "bottom" of the Gulf Stream was probably at between 1,500 and 2,000 meters. Stommel had predicted it would be found at around 1,600 meters. This was no dreamlike idea but an instance, rare in any science at the time, when reasoning before the fact — a hypothesis based on superb physical hunches, on dialogues with colleagues — turned out to be right on the money.

In the thirty years since, Henry Stommel has had plenty more excitement. He has cruised the Kuroshio, the North Pacific's answer to the Gulf Stream; he has studied the recirculation patterns and other phenomena of the Stream; he has investigated motions up, down, and sideways in just about every ocean available, as well as interactions between sea and air. A few years ago he published what he calls "a strange kind of book," the first attempt I have found to invite the layman to consider the fascinations of physical oceanography. The book's title is a reflection of its author: *A View of the Sea*.

Fuglister, Worthington, Stommel, and not an advanced degree among them. Back in the early sixties, Worthington read a report mentioning a new European oceanographic lab being set up, staffed with eight Ph.D.'s and fifteen "subprofessionals." Thus SOSO was born, the Society of Subprofessional Oceanographers.

Stommel was elected president, Fuglister vice president. Worthington, because of his secondary education in Britain, was appointed Ambassador to the Court of St. James. Fuglister died in 1987. Worthington is retired, spending summers in Woods Hole and winters in the Bahamas, where the deep western undercurrent that Stommel proposed and he helped discover "washes my island's eastern boundary fairly closely." Stommel has had bouts with serious illness in recent years, but his geostrophic velocity doesn't seem to have changed much.

Today almost all oceanographers are recognized as such by virtue of their degrees in some branch of oceanography, and many have a knowledge of sophisticated mathematics that floors the oldsters. "I can't do my work with equations," Val Worthington told me. "I have to catch water." The younger man who in effect replaced him mystifies him with his dynamical magic. For his part, Henry Stommel, while apologizing for his mathematical shortcomings, draws the new breed to his theories like flies. He builds the frame, they sheath it with equations.

Although the Gulf Stream isn't quite the magnet it once was at the Oceanographic (a case could be made that more research in that quarter is being conducted at the University of Rhode Island), plenty of Woods Hole oceanographers still work on problems related to the Stream and its structures. Phil Richardson is one. Now in his late forties, Richardson spent years trying to develop information about the current's behavior after it has taken its departure from Hatteras. He studied the departure itself, the vaulting of the Stream over Stommel's deep undercurrent. Floats became his instruments of choice, both surface drifters and later generations of John Swallow's neutrally buoyant device called SOFAR (for Sound Fixing and Ranging). He put his floats into the current at various depths and watched, often in bafflement, as the things went scuttling off in loops or stalled where they shouldn't.

In time the path of the Stream came somewhat clearer to Richardson. His floats showed, as Worthington and Stommel had indicated, that most of the water does recirculate, turning back upstream to rejoin the current. The New England Seamounts appear to deflect the Stream a bit to the southeast and to stir up a good deal of meander activity. But what really drew his attention were the products of meanders, from Hatteras on downstream to the Grand Banks and beyond: the Gulf Stream eddies, or rings.

Overall, the Gulf Stream system reminded Richardson of "an ocean weather map. The Gulf Stream and its eddies are analogous to the atmospheric jet stream and its cut-off high and low pressure cells. The Gulf Stream rings can be thought of as severe oceanic storms, regions where strong currents and temperature and salinity gradients are found." The Stream spawns two kinds of rings. The cold-core variety forms from meanders stretching out toward the Sargasso Sea. If conditions are right, the neck of the meander grows so thin that the main current, seeking the straight and narrow, cuts through it, leaving a whorl up to two hundred or three hundred kilometers across and thousands of meters deep. An old river does approximately the same thing when it cuts through a meander to form an oxbow lake. In their youth, cold-core rings often rotate at speeds of up to four knots in a counterclockwise direction. They contain cool inshore water. Warm-core rings bulge into being on the western or northwestern edge of the Stream. They rotate clockwise and contain Sargasso Sea water. In effect, the two kinds of eddies form a process by which fluid from two vastly different oceanic regimes can intermingle, establishing an important if intermittent interchange of organisms and nutrients. Both varieties tend to drift west or upstream in the countercurrents off the edges of the Stream, and many rejoin it, giving up their energy and water to their parent.

In the mid-seventies Richardson joined a multidisciplinary team from several institutions called the Ring Group, which was charged with tracking as many Gulf Stream cold-core eddies for as long as it could. The team called its best performer Ring Bob. When they sailed into him, Richardson wrote, those on board the research vessel smelled the weedy scent of a summer seashore, and the sea was green and turbid compared to the nearby Stream and Sargasso: "The temperature of the surface water in the center was 15°C, nearly 10° colder than the surrounding water; this difference was reflected in the air temperature." Bob reached right to the bottom, five thousand meters down, perhaps twice as far as the Stream itself.

Eddies have been sensed, if not detected, for centuries. Jonathan Williams, Ben Franklin's grandnephew, found a warm-core Gulf Stream ring at the end of the eighteenth century when he hauled up a cooking pot of water and took its temperature. He called it a whirlpool, but he did associate it with the Stream. It is now evident that the ocean is full of whirlpools — slow, immensely powerful gyrelets that take months to pass a given point while their airborne equivalents, the atmospheric cells, take a few days. Eddy energy is now suspected of driving a good deal more water here and there than was supposed only a generation ago. In fact, it may be so important in the oceanic scheme of things that it will challenge at least part of the geostrophic thinking — the idea of a level of no motion — that has anchored oceanography for so many decades.

Gulf Stream rings are the most powerful eddies studied so far, but they are hard to work with. Their range, the Gulf Stream system, or envelope, is a few kilometers thick, hundreds wide, and thousands long. Satellites work well — sometimes. Floats and other drifters are prone to popping out of the study area. Arrays of moored instruments, like current meters, can measure only what passes their station and can miss a sidewinding current like the Gulf Stream. Shipboard instruments are easier to control, but they can only tell you what is going on at a certain spot at a certain time.

All of which may explain why Phil Richardson has at times deliberately deserted the present. While searching for records that would give a mean picture of surface circulation in the North Atlantic, he heard about the derelicts, the great navy of deserted vessels that littered the world's seas in the time of sail.

That was a killing time; about two thousand ships were lost at sea each year in the last decades of the nineteenth century, taking twelve thousand souls with them. About sixteen hundred turned derelict — abandoned, drifting, some of them half-submerged, direct dangers to shipping. Passing vessels reported their positions, and the information went onto pilot charts.

From that data, Richardson began calculating current patterns. The derelicts concentrated in the Gulf Stream, but some served him better than others. Those with even a trace of top hamper left clearly would be affected by both wind and current, while the hulks barely proud of the water, like Prince Albert's drifters, were the better choice — if choice could be made. The voyages of some of them were in a league with the Flying Dutchman's: the schooner *Fannie E. Wolston*, for example, drifted for three years, making a complete circuit of the subtropical gyre. Vessels that had been in the coasting trade before abandonment usually drifted southwest with the inshore countercurrent, joined the Gulf Stream around Hatteras, and headed northeast. Those not sunk by gales or the explosives of crews sent out to destroy them made the voyage to Europe in about ten months.

The trajectories Richardson got from the derelict data weren't as reliable as those of his surface buoys, but there were more of them. And they backed up his ideas about how parcels of water moved. Although the trackings were made a century ago, "by methods that now seem primitive," he wrote, "I believe they remain one of the best, if not the best, direct measures of the surface circulation of the North Atlantic."

One other venture by this slim and sandy man should be mentioned. It has to do with the Franklin-Folger chart of the Gulf Stream, the first edition, printed in 1769 or 1770 and subsequently lost to history. "It occurred to me," Richardson wrote, "that a copy of the chart might have been saved by the French, because Franklin was envoy to France . . . and because both Franklin and his work were highly regarded by the French." Richardson went to the Bibliothèque Nationale in Paris and talked reticent librarians into letting him look through their charts rather than having them brought to him one by one. It didn't take him long, he said. He had no difficulty in recognizing his find. The legend, the current speeds, the annotations — all agreed with material he had read in Franklin's own correspondence. Without a doubt, Richardson had in his hands the lost portrait of the blue god.

6
Entrained

THE CUSTOM OF CURRENTS is to move. The Gulf Stream system moves water in abundances we know about, thirty million cubic meters of it through the Florida Straits and close to two hundred million cubic meters off the Grand Banks. The water flows in a warm core on the surface and in cool and cold runs below. It curls around on itself and calves into eddies. It swings in larger arcs and returns to itself. It moves rapidly along the envelope of the current and sidles across the sea as the Stream shifts position.

As it moves, the water entrains: sea hearts from Jamaica, trees from some undercut bank along the Mississippi, a glass float from a Portuguese drift net, a broken boat. The Stream railroads a load of tropical and subtropical water up the crossties of the latitudes. And in the load, life floats and swims, moving on siphons and whiplets, on fins that work like wings. In volume, life in the current is dilute compared to what inhabits inshore waters. But it is more various, more eldritch. Caribbean fish show up in the autumnal sea off Martha's Vineyard. Southern turtles ride past New England like sleeping commuters gone way beyond their stop and crawl ashore in alien weather up in the Maritimes or on Scottish cobbles.

Until recently, biologists thought of life in the Gulf Stream as being essentially in transit. The current, they said, was a boundary between the rather unproductive waters of the northern Sargasso and the rich coastal stretches, not a faunal domain in itself. That view is beginning to change. It is still clear that many organisms in the Stream are not there by choice and that many of those will eventually drift north to their deaths. But recent research, particularly biological and chemical studies of structures like the eddies, has brought new dimension to our understanding of their biota. Richard Backus and Peter Wiebe at Woods Hole have shown that although most animals are homebodies and will not cross, say, from the Sargasso and the Gulf Stream to inshore waters and vice versa, there are species that can live in both worlds. Some, the opportunists, seem to thrive where warm water and cold wrap around each other, in the rings. They outcompete other creatures, some of whom resemble them quite closely except for an inability to function in unstable environments.

Backus says that rings make a contribution of sorts to changing the Gulf Stream from a pass-through regime with migrants to a recirculating one with residents. The eddies move upstream, and enough of them recoalesce to produce a circuit. Recirculation time can be quite short in terms of life expectancies: a warm-core ring is pretty old at seven months. Even so, there are species that hatch, live, and die in a given ring. They are annuals, like marigolds. With enough study it might even be shown that the rings have some effect on the evolution of the species they entrain and, further, that the Stream itself, not in its warm core but in the cold water it carries deep in its belly, may be a prime agent in resupplying the northwest North Atlantic with the nutrients necessary to maintain its high productivity as a fishery.

In studying organisms, resident and transient, that support the food chain in and near the Stream, marine biologists have relied until recently on nets. In effect, they have been working in a medium composed of one part organism to one million or even one billion parts water. They have not been effective predators, missing the larval tuna that diving birds catch and the squid that whales capture. From time to time they are accused of being content to net "the weak, the blind, and the unlucky." And the mushable. Many species are so delicate that they become unrecognizable aspic in the meshes.

Enterprising biologists are therefore turning to other means to improve their catch. They work with commercial fishermen and employ commercial techniques aboard their research vessels. And they dive — in submersibles, inside weird pressurized suits, and attached to each other by a Christmas tree of tethers.

Tether diving works well for those studying organisms that are tiny or gelatinous or both. It gives a half-dozen scientists maximum freedom and minimum risk in the open ocean. Without tethers, divers would spread out dangerously far in a few minutes of observing and collecting their animals. With them, each diver can cover about sixty feet, vertically and horizontally, as the whole team drifts with the current. A safety diver keeps an eye out for a colleague in trouble or for the distant glint off a pewter flank that could mean trouble for everyone.

The tether diver I know best is Lawrence P. Madin of the Woods Hole Oceanographic Institution. He is blondish, tallish, with the smooth strength a diver needs. Madin is a salp man, a student of the elegant transparent jars that drift singly and in chains through much of the world ocean.

In the fall of 1985 I flew from Miami to Nassau to meet Madin and his crew aboard the research vessel *Cape Florida*, operated by the oceanographic school at the University of Miami. We were to work along some of his favorite dive sites near the eastern edge of the Stream and in the Sargasso.

We climbed out of Miami, up over a narrow slick a few miles offshore — perhaps the west wall of the Stream, perhaps a trick of wind. Big vessels sailed with the current, fishermen and yachts crossed it, stemmed it, skittered around. After fifteen minutes the indigo turned glass green: the Bahama Banks. Clouds rose in columns through the hot air and sank their templates in the shallows. Sand ridges sharp as hogbacks wandered for miles and belts of emerald water where currents had cut the bottom.

Nassau did not live up to its approaches. The land is flat and scruffy, the town distorted by tourism. The season had just started when I got there. Only three cruise ships were docked, great ferries, all flare and rake, their cargo gone ashore to mill in the shops of Bay Street. In the pink, hot twilight Victoria Regina sat in stone watching the traffic. Noise grew everywhere — radios, monstrous tape decks, the fartings of untuned engines. A man stood in the street talking to his friends in the island lilt. Perhaps five pounds of necklaces and pendants hung on his chest, bare under a designer fatigue jacket. Blacks should wear the gold; it is lost on whites. The man walked up a side street under palms. In an instant their fronds went out of focus, and it was night.

Cape Florida came in the next morning. She is 135 feet long, built more for coastal than for blue-water work. She carries a crew of nine, and twelve scientists. There were eleven in Madin's group, working away at microscopes and aquaria in the wet lab or sprucing up for a few hours ashore. Madin had assigned me to his cabin, the chief scientist's, with a view to port and forward. He said we were scheduled to put to sea late that evening. We'd go north down the eastern edge of the Florida Current and swing over into the Sargasso Sea so that a research team from California could conduct an experiment, then head south up the Stream to Miami, weather permitting — the boilerplate of sailing orders: weather always permitting.

We left the dock around ten at night and lazed through the swells up Northeast Providence Channel, where Columbus might have passed on his way to dubious encounter with the blue god had he not followed his intuition and borne away to the southwest after those migrating land birds. In the morning Madin laid on a practice dive for the few of us who had boarded at Nassau. All the rest were either experienced tether divers or had been certified during the previous week, when *Cape Florida* lay over the vertiginous deeps of Tongue of the Ocean, between Nassau and Andros Island.

I learned to dive thirty years ago in the prime and ignorance of youth, along the reefs of Antigua, just about where *Welcome* made her landfall. My companion and I fooled around at forty feet, hunting conchs for chowder and fritters and trumpets (their shells have been tooted for centuries). Then, in foolhardy fashion, we went a hundred feet down an anchor rope to lie on the bottom at the foot of Submarine Rock. I had done no more diving until I took a scuba course weeks before meeting Madin. I floated and plunged and buddy-breathed up and down an indoor pool and took my open-water work in a hurricane-blasted cove in Rhode Island and on a marvelous wreck deep off the old fishing port of Gloucester.

I had found that diving, like so many other pastimes of the upwardly mobile, is as much business as pleasure, as much cost as benefit. I carried almost a hundred pounds on my body, twelve hundred dollars' worth of equipment, much of which hadn't existed when I went down off Antigua — the buoyancy control device, for instance, a vest inflatable by mouth or from the tank and designed to allow a diver to float where he will in the water column. My mask was set in silicon, my flippers were flukes, my snorkel cost twenty dollars as against two in the golden days.

I was in armor during the Korean War, and I remember that I felt quite peaceful in the turret of an M-47 as it whined and pitched across the red earth of Fort Knox. I felt that way under water in Rhode Island, where I could not see farther than my elbows in the silt clouds of the cove. I found that if I adjusted my buoyancy just right, I could float a few feet above the silt and see: flounders dredged in muck, tautog, northern coral growing on the dingy rock like a disease of the skin, small lobsters behind their sand parapets with their claws up like a boxer sparring. And, deep in a crevice, a tiny remembrance of those glorious southern reefs, a butterfly fish, brought to this stygian coast by the Stream.

Madin had the *Cape Florida* heave to in the channel, well out of sight of land and bottom. The water temperature was a little over 80 degrees, warm enough for a surface swimmer, cool enough for a bit of insulation if you were going down a distance and planned to stay. The divers from Woods Hole used the bottoms of their northern wet suits; I used the top. Madin put on a shorty that protected his torso and upper thighs. We slipped masks and flippers into mesh bags that would hold collecting jars during the dive, and then six of us eased down a rope ladder, an eye on the surges, and stepped into an eighteen-foot Avon, an outboard with a rigid hull and inflatable gunwales.

The bosun took us away from the ship and cut the engine a quarter mile out. Madin doled out collecting jars from a plastic cooler. The safety diver tied the Avon's painter and the tether rig's downline to a fat red float. That done, he clapped hand to mask, wished us well, and rolled over backward into the sea. We could see him in the prisms of the waves, ten feet down, attaching the tethers, mounted on a steel ring, to the downline. One by one we followed him, clipping a tether to lanyards on our gear, swimming off to wait until the full team was tied on. Then we sank together to forty feet, and the scientists fell to work.

They had already debubbled their jars near the surface. Even a small pocket of air left in a jar will make it impossible to open at depth. If you forget about the bubbles and head down, the jar can implode. Odd that, at depth, air is both essential and hostile. The first thing divers learn is to keep breathing. A rapid change in depth with the lungs full of pressurized air and breath held can result in a ruptured lung. Embolisms are quite rare, but like the bends — the surge of nitrogen bubbles in joints and tissues resulting from too much time spent staying down and not enough coming up — they can kill. In fifteen hundred dives, Madin's group had suffered only one accident, an embolism that at first appeared extremely serious. Although the diver lost consciousness when he surfaced, hyperbaric treatment cured him of everything except a slight limp. Oxygen, the best immediate defense against air bubbles in the bloodstream, had not been in the dive boat when that accident happened. It was now, riding in a green box far above us.

The safety diver took us down to sixty feet. I fumbled and tumbled, but gradually what I had come to know won out over what I had forgotten. I found I could use a knot on the downline

as a reference and try for neutral buoyancy in its plane, allowing for its vertical play as the line yanked in the surge, allowing for my own rise and fall to the rhythm of my breathing: intake, a few seconds' lag, then up; exhale, a few seconds' pause, then down. I checked my depth gauge: sixty feet, give or take ten. I will improve, I thought. I am safe. I can look.

Hunting with Madin was like hunting anywhere. If you don't know the telltales you will not see game. I remembered the first woodchuck I shot at. He was feeding at the foot of a Berkshire meadow. My brother, much older than I, pointed, pointed again. I stared along his finger and shook my head. He shoved the butt of his rifle against my shoulder, set my eye behind its telescopic sight, and lined it up with the chuck. Nothing but timothy grass and redruff. Then I wavered past something intensely brown and wavered back to check. There he was, sitting up, looking at me like a chinless old man. Yet I had not seen him — could not have seen him — and missed him when I clutched at the trigger.

There, on the tether, hanging where I never thought to be, I saw blue nothing. After ten minutes I learned to focus in the infinity, and suddenly I had company: jellies, ctenophores probably, pulsing and iridescent and always too far away. When one did swim within my tethered radius, I couldn't capture it. The slightest movement with the jar created the slightest current, enough to keep the organism swirling, slipping over the glass lip and away. Then salps. Turning slowly, I saw a serpent unwinding out of the eternity, three or four meters long, and spotted. I swam out to it, and the spots became brown-gold stomachs in the clear and shimmering flesh of perhaps a hundred salps in line. An open-ended arrangement, this animal. It swims and feeds by forcing water through its tube of a body. Alone, it is an elegant oblong. Attached to its kind, it is a marvel, communicating, cooperating, so that the snake swims. I tugged on Madin's tether to get his attention and pointed to my find. He eased over, unhooked salps from the chain, and they practically scampered into his jars. Ah, well . . .

In half an hour, divers started unclipping from their tethers and heading up the line to the float and over to the Avon. There was *Cape Florida*, maneuvering a safe distance away, protective. We bounced back to mother, stacking jars of collected organisms in the cooler. They would go into aquaria in the lab, some to live on for several days as the biologists watched their respiration, their swimming behavior, their ingestion and excretion. Some, the amphipods and other small bugs, would end up on slides for the microscopes.

Madin met salps when he was an undergraduate at Berkeley. They were small and dead in their jars of formalin, and none of his teachers seemed to know much about them. As a graduate student he spent a year on Bimini, two Bahamian specks perched across the Stream from Miami on the lonely western edge of Great Bahama Bank. Ernest Hemingway used to go there to fish and brawl around, and the passages in *Islands in the Stream* that excite me have to do with fishing off Bimini. Madin would go out to the Stream with the others in a small boat, collect the collectibles, and scamper back to a shoreside lab to work up the specimens. When group members began divvying up the species for specialized study, Madin remembered the little organisms dead in their jars at Berkeley and asked for salps.

Salps are pelagic tunicates. They are related to such strange beings as the tadpolelike appendicularian, which goes about in a glass house. As planktonic herbivores (*plankton* coming from the Greek for "wander") salps are enormously efficient, having much shorter generation times than others of their kind. When they happen upon a patch of bacteria or phytoplankton, their filtering siphons miss very little of it. The salp's ocean is nothing like the marine cornucopia landsmen imagine. Currents, upwellings, and shifts in the basic nutrients of sea life make for a little feasting and much famine. Plankton can bloom or, as happens after extensive grazing of sea plants by tiny animals, fall into blight. The salp adjusts to all save the extremes (a planktonic bloom can clog his feeding mechanism and kill him). When food is plentiful, reproduction speeds up. When it is not, reproduction stops.

And what reproduction! The generations alternate — solitary, colonial, solitary, colonial. A solitary salp reproduces by an asexual budding process that yields a whorl of young. The whorl breaks free of the parent as a chain of females, each with one egg. The egg is fertilized by sperm released in the water by older salps, all of which are males, and develops into a solitary salp, which produces another whorl. Meanwhile the parents, the colonial generation, change sex as they age, thus maintaining the sperm supply.

To biologists, what comes out of the gut is as important as what goes into it. The fecal pellets of salps, which look something like mouse droppings, are many and rich in ammonium, enough to make a difference in satisfying the nitrogen needs of passing phytoplankton. The pellets also sink much more rapidly than those of other planktonic species — on the order of 2,700 meters a day, a fast-food service for the benthos, the community of bottom dwellers. Many salps migrate vertically, the majority upward toward the surface at night and downward — as much as a kilometer — during the day. That appreciably shortens the diurnal fecal trip.

Transparency is an effective defense against predation, but salps are an important food source. A ctenophore or comb jelly called *Lampea*, a gifted contortionist, can ingest salps bigger than it is. Siphonophores, jellyfish, molluscs, and other plankters are daintier but just as deadly. Larger animals, like flying fish and cod, also dine on what must be the leanest of cuisines — salp flesh is ninety-five percent water. So do albatrosses, sea turtles, giant tuna.

In recent years the sea and my back haven't gotten along all that well. After a rough spell of weather, the big muscle on the left of the saddle went into spasm, and I had to settle for one dive a day instead of three. Midafternoon, when salps were apt to be more plentiful, was my spot. I spent the rest of the time reading and going over notes.

I'd just settled into a spell of study when the morning divers came swarming up the ladder from the Avon. They had seen what they all thought was a huge shark. To look at the surface from an appreciable depth is to look through a lens marvelously wrinkled by the swells passing over. Objects seen underwater through a diving mask look larger than life, and objects seen underwater high in that frantic lens look larger still. "He seemed twenty-five feet long to me," said a bearded diver from California who had spent considerable time in the company of sharks, "so I knew he was about eight." As the animal sailed over the team, the safety diver had signaled to the others, not the half-clenched fist with thumb and fingers biting the water — the shark sign — but thumb and forefingers making an imaginary tube growing from his nose. They had been looking, for those few seconds, at a strapping marlin.

I am apt to dance small dances with the unfamiliar, stepping up in curiosity, stepping back in fear. Might see another billfish, I thought on the way out to the afternoon dive. As I sank, the obverse rose, the what-if. Nothing large now, came the whine. Belay the big.

We started hunting at eighty feet. I watched my companions float as a frog floats at the surface of a pond, arms and legs out and at ease. Each time they exhaled, platters of air fluttered up, tipped, and broke into columns of bubbles. A shimmering belt moved below my flippers, a ctenophore called Venus's girdle. Then two individual salps and a winged snail with flesh clear as air. Dots wavered close to me, tiny radiolarians and copepods I could not name in a month of taxonomic study. The only fish in that whole sweep of sea between Stream and Sargasso was less than an inch long. It had fallen in love with my snorkel.

Even less life the next day, when a cloud bank lidded the sun. We swam in purple, and I never reached for a jar. I could have stayed in that royal water for fifty minutes at seventy feet without having to stop for decompression on the way up. But I was bored and went up in twenty, following Madin's assistant, Cathy Cetta. I took my time, passing up my tank and weight belt to the bosun and flippering myself up and over the gunwale. I stripped off my wet suit and let the sun sit on my skin.

"There's a shark in the water." Cetta said it calmly, and calmly

I sat up. There was sudden commotion near the float, the three remaining divers coming on fast, ducking their heads to look down. They came over the side with their eyes very wide, and I could see the shark behind them, swimming close to the fins of the last man in. It was astonishingly brown, as brown as that chinless old man so long ago, with a white-tipped dorsal. About seven feet, discounting the magnification effect of the water. A fishing line streamed down its flank from one corner of its mouth, and I thought of the hook fast in its jaw or corroding away in the acid of its belly. I felt no fear, only wild excitement to be riding there, safe, a few feet from something that could, if it wished, remove appreciable amounts of me.

Tony, the California diver, was laughing, or trying to. "I saw pilot fish," he said. "I knew he must be around." Then he looked at me. "We were worried most about you. The shark came in while you were going up the line. Not much chance he would have taken on the group of us down there. But you looked pretty tempting."

My grin got thin.

"Nothing to worry about," Tony said. "He's just curious. That's just about always the case. If they have other ideas, they don't just swim around you in circles like that. They look a lot more aggressive. Some of them will hunch their shoulders at you."

The bosun started the outboard, and the shark sank away. Plenty of time now for replay. I let my fear rise toward the tempting target of a diver taking his ease beside his boat, handing in his weight belt, shrugging off his buoyancy vest. Lord, let me always sight my sharks when my feet are dry. Never let me look at that beauty, the omega of that mouth, when I'm under.

"It always gives you a rush," Tony said to me, beaming.

It would have been best to go down again later in the day, but my back was tender, and I stayed aboard *Cape Florida*. I found the fear: a large shark had come close to me in the water, and I hadn't seen him. He must have seen or sensed me. Sharks are superbly equipped to know what is around them: they have sight, smell, electroperception. In matters of attack they are the professionals, insultingly superior to a man in the water, to the besuited, betanked alien who swims in open ocean, tied to his kind, blowing bubbles. They can bring not only fear but guilt.

Sharks were to become more familiar. Eighteen months after my introduction, I dove again with Madin and Cetta. We worked in Tongue of the Ocean, in water more than a mile deep, and on the second day we got chased out of the water twice by dusky sharks. In the morning two cut through our formation. One swam right at me, and I suddenly realized that I was retreating from him, leaning backward. When I righted myself, he swung away. In the afternoon a large dusky hung around and refused to leave. We later found out we'd been diving in a field of chum — chopped fish and gurry — spread by fishing boats. The sharks had caught us in their pantry.

Then, as on my first dives with him, Larry Madin treated the predators with respect and some annoyance. They were fouling up his schedule. No shark had ever touched one of his divers, but we would take no chances. From the beginning, we had been taught a defense. When a shark came in, the diver spotting him would alert the safety diver, who would yank on tethers until the entire team had converged, facing outward like so many musk oxen pestered by wolves. The safety diver often carried a shark billy, a plastic rod for prodding, and several of us had metal squares in our collecting bags. A shark will often follow one down and out of sight as it wobbles and drops like a bright leaf; meantime, the musk oxen move up the line and into the boat in a pod. We did that with the duskies. While we were getting into position, one diver unsheathed his knife to cut some line that had wrapped around his gear. The knife got away from him. He felt strange, he said later, to think of it, still in motion those thousands of feet down as our boat bucked its way home through the swells.

On the way in, I thought back to the big dolphins Tony had seen during the *Cape Florida* diving. They had circled around from their cruising to watch the divers, and Tony swam out to the end of his tether to play with them. They hung in the water just feet from him. That night, as Tony was working away at a bowl of conch chowder, he stopped. "You know, when a shark swims at you," he said, making a sideways movement with his arm, "it goes, No, no, no. When a dolphin swims toward you" — slow vertical stroking — "it goes, Yes, yes, yes." Affirmation and negation in the rise of a fluke, the sweep of a tail.

It is perfectly possible that during the dives with Madin I could have swum right past a leptocephalus or two. Not knowing them, I would have looked right through them. They are small, smaller than my little finger, and transparent, the shape of the beech leaves that swirl and sink in the dark clarity of the Deerfield River late in the fall. Once thought to be a separate species, leptocephalus — slender-head — is now taxonomically fixed as the larval stage of *Anguilla*, the eel.

I wish I had seen a lepto and recognized it as I moved, entrained with it, on my tether. The leaflets are the only plentiful evidence we have of *Anguilla*'s life in the open sea, a life whose principal passage is entrainment in the Gulf Stream system. A handful of adults have been captured on the continental shelves of eastern North America and western Europe. One has been photographed in blue water but none have been taken there, nor have any eel eggs been identified in plankton hauls. *Anguilla* is a mystery, something that most delights us about the sea. Increasingly, with the eel and so many other near-fantasies, biologists no longer act like biologists. They cannot work directly with their subjects, so they infer and simulate, the way physicists do. They proceed by learning what the physical oceanographers can tell them about how the ocean moves and when.

Those interested in eels used to proceed by imagination. What do you say about a being that disappears from your estuaries as a silver snake and returns as a glass worm? You say that it comes from the entrails of the earth, that it reproduces by sloughing off bits of skin on rocky shallows. In Sicily shortly after the birth of Christ, some said that the female eel mated with a snake of the land, who first paused to empty himself of his poison and then hastened to the beach, where he delivered himself of a whistling love song. Hearing that, the eel threw herself on her lover, popped his head in her mouth, and *kaboom!*

In the early 1900s, a Danish biologist named Johannes Schmidt began collecting eel larvae in the eastern North Atlantic. Eels are an important food in the lowlands and elsewhere in Europe, and Schmidt's studies, like so many other oceanographic ventures of the time, were geared to the commercial catch. Over a couple of decades, he fished farther to the west and south, looking for the area where his nets caught the smallest leptos. He found it, in the Sargasso Sea, east of the Bahamas. This, he said, was where *Anguilla* came to spawn. The larvae of *Anguilla anguilla*, the eel of European rivers, took two and a half or three years to get back home. The American lepto, *Anguilla rostrata,* took one.

Biologists have tested Schmidt's theory and found it sound enough to build on. Once upon a cycle, my translation of that theory goes, there was a large yellow-bellied eel in North Dakota or western Pennsylvania or Lake Ontario or the Passamaquoddy or, in my fancy, the Deerfield. She was well over two feet long and between five and ten years old, and she went about her business of catching frogs and water rats and fish and spending her winters in her favorite muds. Old Yellow had a purple eye, an army-issue back, one long fin that ran down her spine, around the tip of her tail and up to her anus, and a respectable set of pectorals.

Fall came on, and Old Yellow began to change. Her back turned dark, almost black, and her sides and belly the color of moonlight. Her nostrils widened, her eyes doubled or more in size. Her sexual organs swelled and her digestive system dwindled, and she was ready.

Sea runs are common things. Trout and salmon and shad and many more animals spend time in the salt and return to the rivers to spawn. They are anadromous. Eels are the rare ones, catadromous, river dwellers and ocean spawners. I see what was once Old Yellow, silvered now, late on an October night, leaving her place near the old flooded quarry a few miles upstream from where the Deerfield flows into the Connecticut River. Down she swims, past the frosting bottomlands, under the old railroad bridge, into the mother river, and south to the sea. She lies up in the bright days and moves in the dark, and around her other eels move. Past the opera house near Haddam and down to tidewater she goes. With the taste of salt come the males, smaller, hangers-out in estuaries and in marshes. The eel from the Deerfield waits in a dense concentration of her kind, and at last the signals come — a new response to the ebb of tide, to an inner shift of chemicals — and the mob runs over the bar and into Long Island Sound and out with the tide onto the continental shelf. Out across the slow deepening to the shelf break, out into slope waters with the continent falling away beneath them, out to inference.

Some biologists have thought that eels pitch down into the abyss, perhaps following a deep countercurrent of the Stream to the Sargasso spawning ground. But James McCleave of the University of Maine at Orono discounts that. McCleave, one of the most active apodologists in the United States, argues that in metamorphosing from yellow to silver, the eel aims at migration in the upper layers of the sea, where at least some light penetrates. The retina adapts so that it can see well in the blues of the euphotic zone. Changes in the swim bladder are such that the bladder can't remain fully inflated much below 150 meters.

But McCleave has what he calls a "nagging" problem: a photograph of an obviously gravid, healthy *Anguilla*, taken by *Alvin*. Woods Hole's little submersible was driving across the bottom of Tongue of the Ocean when she came upon the eel, or vice versa. The location wasn't very far from where Madin and his divers like to collect specimens, and the depth was just over 2,000 meters. Bothersome, but McCleave sticks to his euphotic guns.

It has been discovered that eels, like many other migratory animals, have bits of magnetite in their heads. Perhaps that enables them to set course for their spawning. They do not need to hold to a strict heading, since a generally southerly course will get American eels where they want to be, while a generally southwestern one will suffice for the Europeans.

The two spawning areas overlap, the Europeans preferring waters a bit to the east of those favored by the Americans. Both may rely on a physical signal to tell them when to stop swimming and begin shedding eggs and milt. In the winter and early spring, the eel spawning season, a front stretches east and west in the southwestern Sargasso, cooler and less saline water to the north, warmer and saltier to the south. McCleave and others believe that this front, along with the amino acids and other chemicals that tend to concentrate in such structures, may provide the essential triggering cue.

The whole thing seems awfully inefficient — a swim of a couple of thousand miles for the Americans and more than three thousand for the Europeans. But the eels are ancient animals. It is perfectly possible that they were spawning in this same spot more than 150 million years ago, when the North Atlantic was little more than a rift. With continued spreading of the sea floor, the spot moved to its present location, and the eels followed.

American and European eels are almost indistinguishable; the Europeans have a few more vertebrae. They intermingle on the spawning ground, and so do their spawn. The eggs hatch quickly, perhaps within two days, and leptocephali in their millions of millions, each less than ten millimeters long, rise slowly from the depths and begin their own migrations. Where the currents go, they go.

A few leptos have been found in the Caribbean, an indication that there may be some spawning there. These most southern of the American larvae are carried west and then north through the Yucatán Channel, where the cutter *Dauntless* lay in wait for drug boats, and into the Gulf of Mexico. Some of them, as well as those leptos that have drifted through the Bahamian passages, end up in the Straits of Florida. But the mass of the American eels appears to drift slowly northwest, forming a feeder patch just north of the Bahamas, which delivers leptos to the Gulf Stream in such a way that their distribution is fairly even throughout the current, from Hatteras up past the Maritimes.

The European larvae, starting farther east than the Americans, appear to disperse slowly throughout the Sargasso as it wheels with its slow, clockwise circulation.

On both sides of the North Atlantic, the approach to the home coast involves another triggering cue. The leptocephali now metamorphose from leaf to cylinder. A few spots of pigmentation appear, mostly along the spine, but the animals remain largely transparent. They are now glass eels, elvers, a few inches long and ready for the river.

The cycle is enormously simplified here. Ocean currents are complex, stochastic, chancy. Most larvae never make it to metamorphosis — nor can they, if ecological balances are to be maintained. They hatch in astronomical numbers and they perish in them, carried away into the ocean interior or the mortal chill of high latitudes, grazed upon as one plankton patch encounters another.

The miracle in the western North Atlantic is that the new generation of *Anguilla rostrata* ever makes it ashore. They have a Gulf Stream to cross, and they must do it at just the right time. It is possible that some get carried across in eddies or meanders or Gulf Stream intrusions onto the continental shelf, but these phenomena represent such a small percentage of total Stream transport that they probably aren't all that important. Henry Stommel and others have shown that other forms of cross-Stream movements exist, but these appear to occur mostly below 500 meters, whereas the leptos tend to be found in the upper 150 meters at night and 350 meters during the day.

No one knows just where leptocephali change into glass eels. James McCleave is exploring the idea that metamorphosis occurs not in the open sea but when the leptos — what is left of them — encounter the continental shelf. How they get there is the question. Sometimes the sea can be helpful. The increased turbulence in the Stream near the New England Seamounts might make it somewhat easier for the little animals to close on the coast, and that may be why so many elvers are found in river mouths nearest the Seamounts.

Whatever the mechanism of arrival, the journey has got to be a rough one. The swimming abilities of the elvers are nothing compared to those of Old Yellow. The tides could be a problem, but the elvers seem to know how to deal with them. Coming toward land, they drop toward the bottom on the ebb and rise into the water column on the flood. Perhaps a change in noise level, perhaps a different odor, a difference in electrical cues as the water changes its course through the earth's magnetic field, tell the immigrants what they need to know.

In New England the glass eels begin coming back into the rivers in late winter and early spring, five seasons into the cycle. They are not like the salmon. McCleave and his colleagues are fairly certain that they are not returning to natal water to spawn. No offspring of Old Yellow is likely to push up the flooding Connecticut and the Deerfield to the quarry. For them, any decent stream will do. Arrival in sweet water is enough.

Currents are invasive, bringing water masses of different temperatures and chemical compositions into contact. On land, biologists recognize an edge effect, a tendency for life to concentrate near boundaries between pasture and woodland, marsh and prairie, other biomes. At sea, anything with edges attracts. You can probably find a fish of some sort under a floating broomstick. Where different waters converge, water is apt to sink, drawing in anything from copepods to long windrows of weeds. Students of sea turtles have recently discovered that the young of several species spend time in and around those windrows, especially those near the Stream. (It now appears that some adult turtles may ride the Stream in a migration that eventually takes them completely around the gyre. The waifs occasionally found on British beaches are, according to this notion, not there purely by chance but because they took the wrong turn and followed the North Atlantic Current into waters whose temperature they cannot tolerate.)

Edges of grander scale, like the margins of the Stream and the rims of its eddies, are usually rich in life. Plankton, eggs, and larvae collect there, and the fish that feed on them, and the fish that feed on the feeders. Warm-core rings, moving southwest inshore of the Stream, are favorite spots for commercial fishermen, who like to fish close to where the envelope of Stream water shoulders the coastal water aside. Big billfish — marlin and swordfish — swim there, and the bluefin tuna.

Of all the fish associated with the Gulf Stream system, none is more remarkable than the bluefin. Be one in your mind. You course through the Straits of Florida, sweeping along the edge of the Bahama Banks. You may feed here and there, taking squid or one of the mackerels, flying fish, even salps. You are heading down the Stream for the fattening grounds off New England and the Maritimes. When you have fed, you will weigh about eight hundred pounds, two hundred more than you do now. You are a good-sized giant. The record for a bluefin taken by a sporting boat is just under fifteen hundred pounds. And it may be that somewhere the king lives, almost forty years to your twenty, a ton of some of the most highly specialized flesh on the planet.

Your name has the sound of your body shouldering up the sea

when you close on prey. *Thunnus thynnus*, it goes. *Thunnus*, from the Greek for rushing. The Cubans who fish for you call you *atún de aleta azul*, the Norwegians *thunfisk*, the British *tunny*, the Portuguese, who have known you long, *atum*. Your meat is not white like that of the albacore or light like that of the yellowfin tuna and the smaller of your own kind, the meat the packers sell to the American public as sea fowl. Your flesh is mostly the color of darkest blood, and it tastes of lemon and strength.

You are male, recently come from spawning. Earlier in the year females dropped clouds of eggs, perhaps ten million apiece for the larger among them, in the Gulf of Mexico and in the Straits of Florida. Some of the eggs you fertilized will survive to become larvae. Some of the larvae will survive to become infants, "zero" fish, that will work their way into the Florida Current and up to the coastal waters off the Carolinas and farther downstream. Fish that reach a year will weigh eight pounds on average. Five-year-olds will weigh more than a hundred. Ten-year-olds will be around three hundred pounds, the lower limit for giants. As long as you live, you never stop growing.

Look at you. A great mackerel, some might say. You are of the family, a scombrid. Up in the Maritimes and off Maine, they used to call you horse mackerel. Your skin, so sensitive you shudder when touched, is countershaded like that of all scombrids and most other fishes that spend time in sunlit water. Your back is dark, near black, glinting gray and green. Seen from above, you are part of the dusk below. Your belly and sides are silver, dull to polished, sometimes with a flush of pink. Seen from below, you are part of the wild light under the bellies of the seas.

At your size, you don't really need the camouflage to hide from predators. A big mako might take you; he might come in on you when you are feeding heavily, speeding through the baitfish and severing the muscles of the peduncle that control your tail. Then he could finish you as he wished. Sometimes he handles big billfish that way. The animals you watch for are some of the toothed whales — the killer whale and the false killer. They can kill you directly.

Most fish are neutrally buoyant, equipped with gas-filled swim bladders that enable them to hang at a given depth, conserving energy until they need it. Your bladder system doesn't have that buoyancy. You are heavier than water, so you must fly in it. Some other fish, like the great white shark, do this too, but sharks are fixed-wing aircraft; their pectorals remain rigid. You have swept-wing design. You sight prey and become a bullet, your first dorsal fin, the triangular one, depressing into a slot in your back and your pectorals folding into grooves in your flanks. Your fusiform body stays rigid. Your new-moon tail, one of the most efficient of living propellers, picks up the beat. Only when you are near your target do your pectorals and first dorsal extend, giving you the control you need to feed well. At the last moment your gill covers pop open, and the extra suction draws in bait from the sides as well as in front of you.

You can sprint at speeds of fifty miles an hour or more. Some physiologists infer that you might go faster, so fast that without other mitigants in the equation the muscle heat you produce should be enough to cook you. But even at rest you cannot be restful. If you stop swimming, you drown. Your ventilation is of the ram-jet variety, not the pulsing flow across the gills produced by slower fishes. You must travel, at over three knots at a minimum, to breathe efficiently.

In the last couple of decades, scientists have learned something of bluefin physiology. A couple of these researchers work at the Oceanographic Institution in Woods Hole. John Teal is a marine ecologist whose interests have run from salt marshes to the Sargasso Sea. Frank Carey is more specialized, a biologist dedicated to the behavior of certain pelagics — bluefin, swordfish, sharks. He has enormous swimming strength himself and has dragged many a tethered friend along, fuming helplessly in his mask, while he pursued something interesting. Teal and Carey have discovered how tuna, and most particularly the bluefin, manage to perform their locomotory feats.

That tuna are "hot" is common knowledge among biologists. What Carey and Teal did, back in the sixties and seventies, was show how heat was conserved in the body. The mechanism is a countercurrent heat exchanger which, in bluefin, runs down each flank close to the lateral line. The system works somewhat like a refrigerator in reverse. Nets of closely packed veins and arteries function so that the heat from venous blood leaving the muscles is transferred to cold, oxygenated arterial blood coming in the opposite direction, from the gills. The result is that the bluefin can maintain advantageous temperatures in its major muscles while swimming not only in tropical but in subpolar waters. Inside, the fish can be a sunny 86 degrees Fahrenheit when the waters outside it are below 50. In fact, so well does its furnace function that it may have to be careful to damp it down. After a series of sprints, bluefin are known to head for deeper water, quite possibly to cool off. Hot muscles function better than cold ones. Nerve impulses are transmitted faster through warm tissue, and the muscles themselves can contract and relax faster. This is part of what is known as metabolism modulation. Tunas are good at it, and so are bumblebees.

Heat exchangers, large hearts and blood volumes, respiratory surfaces almost equaling those found in mammals, one of the greatest concentrations of muscular hemoglobin found in any animal — the bluefin clearly has been working on something. Marine ecologists describe that something as an expansion of range. Tuna started out in the tropics, but they elected for nomadism. They gave up the risks of residency in one patch of sea — a dwindling of food supplies, for example — for the opportunities of capitalizing on food abundances across whole ocean basins. If they could move easily vertically and horizontally in the water column, if they could cover thousands of miles and survive, they stood a good chance of finding food. They could also feed on it with less competition than they would find were they to remain resident in a small region. The fact that food is patchy has not kept them from the successful completion of their rounds.

The man who has done more to draw attention to the marvels of tuna migration than just about anyone else is Frank Mather.

Mather started out as a naval architect, and right after World War II, when he was visiting his family's summer home near Woods Hole, he hired on at the Oceanographic. He had no doctorate, only a knowledge of boats and a developing passion for big-game fishing, and that's where he started. He went to the library and researched what there was to research about the tunas, marlin, sailfish. He worked on them in the lab and he fished for them, increasingly, up and down the coast. He was a fish catcher, he said, and his oceanographer friends were water catchers. "Catching water is easy," was his motto. "Catching fish is hard."

Mather developed a system for tagging pelagics with small darts and then releasing them. The darts carried messages asking those who recaptured the fish to send information on date, location, weight, and the like. He also began the difficult job — it would be next to impossible now — of trying to convince sport and commercial fishermen around the Atlantic to actually let an occasional pelagic *go.* That way, he could tell something about the migration patterns and other habits of the species.

The problems of his program gave Mather some special things to hate, and his language beats any I've heard on trawlers or scallopers. It is more imaginative, more intense. But in his sack of invectives no term is honed to a greater edge of insult than "administrator." "If you want to fuck something up," he says, "get one of those."

In the Atlantic, bluefin have been taken off Europe, Africa, and the Americas. Columbus reported sighting them. What Mather's tagging program helped to show was that bluefin are true tourists and that they often use currents like the Gulf Stream on their tours. Migratory patterns came clearer on the western side of the North Atlantic, where tagging has been more extensive. Mather was able to show that fish recently spawned in the Gulf of Mexico, the Straits of Florida, and, possibly, adjacent areas, tend to move down the Stream to summer feeding areas in coastal reaches between Hatteras and Cape Cod. In summer the younger fish move out to the deeper and calmer waters over the continental shelf nearby. The larger the fish, the farther north it swims — into the Gulf of Maine and off Nova Scotia. In winter the giants appear to head south, dispersing from Bermuda into the equatorial currents and beyond.

The situation on the eastern side of the Atlantic is more complex, but if you think of the Gulf of Mexico and the adjacent Caribbean as the American Mediterranean, similarities become more apparent. The Mediterranean is a major spawning ground. From there, the little "zero" fish swim out to gather near the Moroccan coast and in the Bay of Biscay. As they grow they continue to move north in the summer and drop down south, somewhere near the Canary Islands, in the winter. The giants make their regal arrival in the Mediterranean in May or June, spawn, then return to the open sea, some of them pushing north past the British Isles to summer off Norway.

Nothing can be more hypothetical than charting the migration of powerful animals of varying ages and habits through a medium that is itself not well understood. Some brave students of bluefin movements in the North Atlantic have tried for a unifying theory, the unifier being the subtropical gyre: some giant fish are said to use it, leaving the Gulf of Mexico, passing along North America and striking over to Europe, then returning via the equatorial current. Neither Mather nor the people at the National Marine Fisheries Service lab in Miami who now operate the tagging program put much stock in that. Some small bluefins are known to have crossed from west to east, from the nurseries of Hatteras–Cape Cod to the nurseries of Biscay. These migrants, however, belonged to only a few year classes (groups of similar age), and they crossed only in certain clusters of years and only in the cool seasons. Mather thinks the urge to cross might be related to the stimuli that cause the fish to head east before winter to deep water; some just keep going.

Some giants also cross, but they appear to do so in the warm season. Most were tagged in waters fairly close to the Stream (when Mather was keeping records, no giant tagged in coastal waters and released subsequently went transatlantic). The big fish generally move out east of the Grand Banks and along the course of the North Atlantic and Norwegian currents to Norway. Most of those recaptured there have been thin — the Norwegians call them long-tailed tuna. One researcher checked the cycles of these trips and discovered that the peaks occurred during periods of strong westerly winds; an extra reason, perhaps, for catching a ride on the currents.

Scientist and bluefin are going to have to get more intimate if the clues Mather and his colleagues have gathered are going to be turned into solid models. The gyre does seem to influence at least some migratory patterns, but the relationship is as little understood with tuna as it is with eel, with turtles, with animals we don't yet even recognize as riders. A fish in a current cannot sense that current as a hawk senses his location in a thermal. He must have different ways to find his path in the sea. There is evidence that the bluefin may practice some form of celestial navigation. Biologists have discovered a pineal window in their heads, an unpigmented patch of skin below which lie neural arrangements connected to the brain. Light passing through the window may generate orientational signals. After all, if horseshoe crabs have a rudimentary celestial eye, why not the great wanderer?

It is said that the tuna are the most hunted of any wild animal. Certainly they have been hunted long. Millennia ago, shore dwellers in the Mediterranean and beyond its mouth were building traps to catch bluefin going or coming, and fortunes were made in the flesh. Maine Indians harpooned bluefin six thousand years ago. Five thousand nine hundred years later, American fishermen took them now and then for oil: a strapping giant might yield twenty-four gallons of it. Real fishing pressure didn't develop until boats and gear came along that were versatile and seaworthy enough to intercept the nomads far from the western shore — gear like the purse seine and the longline. The Japanese have been long-lining for centuries, though they didn't concentrate in the North Atlantic until fairly recently. They are the champions of the art. Twenty years ago, when distant-water fishing was at its peak, they set twelve million miles of lines in the world oceans. Along each mile were thousands of short baited lines. The sets drifted deep, dangling from buoys, and they caught pelagics by the hundreds of thousands of tons each year. Purse seining is a favorite among tuna fishermen. They can set on a school of fish, drawing the long mesh around it and then closing the bottom with the equivalent of a puckering string. It is hard to tell which method is more efficient, the net or the line, but it is safe to say that overuse of both has done serious damage to bluefin populations.

Sport fishing for big bluefin officially began a century ago, when an angler took a fish of more than six hundred pounds off Nova Scotia. I read when I was young about Bluefin Boulevard in Soldier's Rip, near Wedgeport, Nova Scotia, and about the bluefin tournaments there, graced by rich men with huge reels built like Swiss watches and rods like saplings and names like Kip Farrington and Mike Lerner. Rich men, rich in money and time, competing for the giant tuna feeding in the rip. Men from around the world, remembered in what now seem silly salutes: one was honored as "a great English sportsman, citizen, soldier and stockbroker."

I read of the tournament at Cat Cay in the Bahamas, of how tens of thousands of pounds of bluefin were brought in to be measured and weighed. Some fish were dumped at sea, some buried in the golf course to improve the color of the greens. That didn't bother me then. I loved fishing and hunting and didn't think to put the true price on their excesses. I read the books of a dentist named Pearl Grey who changed his first name to Zane and his profession to chronicler of the outdoors. I read Hemingway and learned what he felt about taking marlin and tuna in the headwaters of the Stream as they flowed past Cuba. I was in love with his images then. I am in love with his images still, reading those few paragraphs in *Islands in the Stream* in which the big hammerhead comes over the reef at the boy spearfishing in the shallows.

I'm not so sure about the man himself. There is a sad story

involving bluefin off Bimini. Hemingway evidently caught a giant and drank seriously to his success. When someone at the dock disparaged the catch, Hemingway came rushing out, roaring. A bystander said she saw him in the last light, crouching by his fish as it hung from the scales. He was hitting it, over and over, punch after punch.

When I was first talking to Frank Mather about bluefin, I asked him if they struck hard. He had a conference on tuna management to get to in Washington, and I was driving him to the Boston airport in my pickup. "Do they hit hard?" He repeated the question as if he hadn't heard it, and then he laughed and cackled. "Oh, Jesus! It's like a speeding car dropped into the water. It's like you dropped an anchor out your window right now." Bluefin spurt for the bait as they do for any food, he said. They jump on it sometimes. Sometimes when they're hooked they run on the surface, "making a roostertail like a speedboat, higher than this car."

I went out with Mather after bluefin, in Massachusetts Bay in the summer of 1985. He had a young doctor friend with a powerful boat equipped with a tuna tower and a pulpit for harpooning. The doctor decided to try chumming first, anchoring on Stellwagen Bank in a tangle of craft ranging from lobstermen to the sleekest sports fishermen. The idea is to present visiting bluefin with choices of chopped fish at varying depths, oils and shreds of flesh spreading downstream to lure them in, a more concentrated gurry than the chum I encountered diving with Larry Madin in Tongue of the Ocean. Depth recorders spotted bluefin below us, each leaving a mark like a tiny fir tree on the screen. Mather said a few of them were full giants. None of them took the baits drifting so artfully with the chum.

In the afternoon we trolled. By then I had the routine down pat: the greenhorn can hold something if he is asked to do so, but on no condition may he help. If he helps, chances are he harms. I held Mather's hook as he tied a Bimini twist knot, and I held his rod while he checked this and that. He was fishing 50-pound test, six hundred yards of it. The record for that line is a fish of some eight hundred pounds. It must have been an excruciating fight. You can use the fighting chair, and that helps a lot because you're strapped to the rod by a wide harness belt. But with line that light, you can't ride the harness. I learned that with more traditional lines, like 130-pound test, the angler uses his legs to whip the fish. He jacks himself up off the chair with his legs and then sinks back, raising the rod without overly tiring his arms. One of the best giant killers is an aging woman. She trains for her fights by riding bicycles to get her legs in shape.

To keep from trying to help Mather, or any other fisherman in the cockpit, I climbed high to the tuna tower. Perhaps I could see signs of bluefin feeding. When the fish are working on a nearly stationary school, they can turn the water nervous. It jitters over their workplace. And when they are chasing a school, they can make one collective wake, a humped vee of water marking their sprint. There was nothing. No sign. And then four tuna — mediums from what I could see — destroyed the surface, smashing and turning. Four seconds, maybe five, and they were gone. I watched the sea smooth over.

No market for those four yet, but a market there is. The Japanese love raw fish. Their beluga is giant bluefin, fall fish, at their fattest, and they pay more than fifty dollars for prime belly meat in the better restaurants. They catch all they can and then buy from foreigners to meet the demand. Japanese agents and "buy boats" are on hand near Montauk and Gloucester and other easily accessible ports, and they are willing to pay an angler or a boat captain several thousand dollars for a prime giant. The fish are packed with ice and antibiotics inside plastic coffins and flown to Japan. That traffic is turning many a sports angler off New England and the Maritimes into a commercial fisherman.

For at least a couple of decades, such fishing pressure has put the bluefin at great risk, and that has helped keep Frank Mather mad as hell. Estimating the demographics of fish may be terribly hard work, but it is harder still, once the work shows some halfway acceptable findings, to persuade the fishing community that they have meaning. A healthy population includes fish from many spawnings. What Mather and others began to see back in the sixties was something like what is happening in the United States today, a shrinkage of the middle class. Fishermen were taking so many smaller fish that whole generations were all but missing, and, another danger sign, there were many more giants taken. As a stock shrinks there is less competition, more chance for rapid growth. It got so it was easier to catch a seven-hundred-pound bluefin than a seventeen-pound bluefin.

Mather wrote papers, talked to influential people, even agreed to go among the administrators, to participate in the deliberations of an international body set up to try to stop the decimation of the tuna. What had seemed a likely demise of a species has slowly been worked around to a situation that is precarious but manageable. Fishermen are now accepting limits on their pursuits. The Japanese are even working on culturing bluefin in the Pacific, raising them in ocean pens for the market. One man alone couldn't have brought that about, couldn't have come close. But one angry man surely made a difference.

7
Proxies

IF I AM to have an old age, I wish to live it as Benjamin Franklin did. Oh, he might have been given to calling himself a fag end, a remnant of what he once was. But think of him, a man of seventy-nine suffering from gout, the "stone," swollen legs, being carried in triumph on a litter as he left France for his homeland. When his hosts finally let him depart, he crossed the Channel in composure while the other passengers were turning white and sprinting for the rail. In England he boarded the London packet with his grandnephew, Jonathan Williams, and during six long weeks at sea wrote his wonderful "Maritime Observations." With Williams's help he also recorded the sea temperature every day, at eight in the morning and six in the evening, and kept a log of everything encountered, from gulfweed to bioluminescence.

I imagine that Franklin's run of the ship during his transatlantic crossings increased in direct proportion to his fame and age. With that golden curiosity, he certainly came to know his vessels and their shortcomings. His suggested improvements, from sails to soup bowls, went on and on, yet I have been unable to locate much in his writings having to do with ships as platforms for his science. We know that just before the Revolution, he crossed to France aboard the *Reprisal*, an armed American sloop, and that he had much to do with merchantmen and mail ships before and after that. *Sloop* and *packet* are words that give no exact idea of length or rigging, but it is probable that the London packet of

Franklin's last voyage was ship-rigged, a three-master roughly ninety feet long on deck.

Late in life Franklin had the wherewithal to take a full cabin for himself, "that I might not be intruded on by any accidental disagreeable company." There, he could write and muse on his readings of temperatures from the surface down to more than a hundred feet. But in making those measurements, Franklin was bound by the master's decisions on courses to be run. He was always a passenger, and a caring one, concerned even for the luckless in steerage. "It is not always in your power," he wrote to would-be sea travelers, "to make a choice in your captain, though much of your personal comfort in the passage may depend on his personal character, as you must for so long a time be confined to his company, and under his direction." There are good-natured masters, but if the one at hand "happens to be otherwise, and is only skilful, careful, watchful and active in the conduct of his ship, excuse the rest, for these are essentials."

The scientist at sea today may have only a bit more freedom in choosing his skipper. But in all else he is a different fish, neither passenger nor crew. He (the number of women oceanographers is still small, though growing) is aboard to work. He stands watches and works under the general guidance of the boss of his party, the chief scientist. Still, except in conditions of high risk for the ship and her complement, his real master is his project. The crew and its captain are responsible only for getting him where he wants to go, maintaining him there in as much comfort as can be mustered, and depositing him back on the dock.

That is much more difficult than it sounds. Operators of the twenty-five vessels in the American research fleet need to get as much sea time from them as they can. Ship schedules are drawn up generally well over a year in advance. Research voyages or cruises are shorter than they used to be — or as they still are in countries like the Soviet Union, whose ships are gone for months at a crack. Coastal vessels normally go out for a week or two, and blue-water ships for three weeks or a month, depending on the work to be done. Often, cruises are divided into legs separated by a day or two in port somewhere to take on new equipment, crewmen, and scientists.

Rising prices, especially for diesel fuel, and rising complexity of instrumentation have nicely stressed the process. Forty years ago Columbus Iselin hollered about his "A-boat" costing $400 a day. The cost for the *Knorr*, the Oceanographic's largest vessel, runs to about $12,500 a day, exclusive of scientists' salaries and expenses. So the planners pare and piggyback. There is more space to play with than there was decades ago, when scientists often had to "hot-bunk it" — one clambering in under covers left warm by another recently gone on watch. But today there is also much more equipment to go into that space: banks of computers and electronic gear, miles of special cables, chains and line for towing instruments or stringing them together in moored arrays or lowering and raising them. On a given cruise one research project may require the ship to spend time in one piece of ocean, while another may call for days of constant steaming. Even Frank Mather would have sympathy for the poor administrator faced with reconciling such demands and soothing such serious egos.

The chief scientist keeps an eye on the work of reading the sea, and this job is rarely assigned to the inexperienced. I once spent some time in the wilds of Georges Bank with a young physical oceanographer who never seemed to sleep. He was everywhere, suggesting this, calling for that, trying to get some huge, instrumented tripods up off the bottom, where they had spent months gathering data and beards of sea life, and replacing them with a new set. And all this against a deadline set by the approach of a hulking northeaster. As the wind built to a yodel, he turned away from the rail, put his sopping face next to my ear, and yelled, "I wonder if there'll ever be an easier way to do this."

Not likely. There are a lot of things you don't have to go to sea to do. You can work and rework information. You can run laboratory experiments. You can teach. But in the long run, someone still has to get wet. The best oceanographers either love going to sea or they do it to preserve status or to assure themselves of particularly tasty data. Many spend a couple of months a year out there, some more. Personality traits have a lot to do with their success. It is no accident that sailors have survived so well as prisoners of war. Samuel Johnson was right: Going to sea is a bit like going to jail. Those who perform well adapt to the incongruity of bobbing around on the top of all that alien fluid. They perform their tasks, stand their watches, and spend their free time tinkering, reading, sleeping, or lost in a special reverie.

Time and again, in the first heavy weather of a cruise, I've heard men moan the equivalent of "What am I doing here . . . again?" It is hard to concentrate aboard ship. The four-hour watches neatly derail trains of thought, and a growing exhaustion keeps them off the track. Sounds distract: the pneumatic paint-chippers hammer like obscene woodpeckers, and even the best-tended machinery in time begins to cha-cha. In a storm, there is no mercy to the motion. I've heard that one man on the old *Atlantis* considered taking sleeping pills, but desisted when he realized that if he ever did fall asleep, he might be thrown out of his bunk and killed.

I have been advised by the deans of the seagoing not to believe anyone who says he or she never gets seasick. There are ways to mask it. On *Welcome*, I took to wearing a little drug-saturated disc that, placed behind the ear, lulls the inner ear. I found it also turned me into a zombie, and after walking into the foremast twice of a breezy afternoon, I ripped it off, went and found a bucket, and soon returned to the domain of the living. There is folk art in ways to keep nausea at bay. Wear a wrist strap with a little node on it that will press into your flesh just so. Or put a strip of tape across your bellybutton and face into the wind. No. For me, the best thing is to embrace the beast, fire off a round to lee, and wait for things to get better. For most of us they do. It is when the thing becomes chronic that real danger — dehydration and other nastiness — comes into play.

True oceanographic skill lies in transcending the state of sea, mind, and belly and getting good work done. Much of that takes place on the fantail, the long, low deck astern where the proxies, the substitutes for human senses, lie ready: current meters, long tubes that measure horizontal or vertical flows past them and record those measurements on tape; buoys shaped like giant doughnuts, instrumented to read wind speeds and directions and the physiology of waves; floats that can be persuaded to follow a certain parcel of water or to drift along at a certain depth. Some of these things weigh a ton or more, and their handlers are enormously talented people. Woods Hole is supposed to have one of the finest buoy groups in the world. I watched them work once during several weeks of multinational operations west of the Hebrides. They had put weeks of planning into what they would do. Collectively they had more than a century of experience in lifting, steadying, and putting loads into the sea right where they were supposed to go. They used air-tuggers — small winches — and frapping lines and tagging lines. And they constantly checked themselves and each other to make sure they stood free of the bight. A line running from a cleat to a buoy and back to a winch is a slingshot. If you're inside the angle — the bight — and something gives, you may too.

Research vessel design was coming along well some decades ago, and the Oceanographic's R.V. *Knorr* is a fine example of that advance. She has cycloid propellers, vertical and adjustable blades set in a rotating base. With them, and with navigational data from a passing satellite, she can maintain position in any patch of sea. That is an immense advantage when you're investigating something on the bottom a couple of miles below your keel — something, say, like the wreck of the *Titanic*, which *Knorr* helped to find. But the ship is twenty years old. And she is limited, as are all of her sisters, by weather.

"Forty knots of wind at sea," says Allyn Vine, "will probably slow down scientific programs about as much as a foot of snow in Washington." Vine, retired from the Oceanographic, is an establishment unto himself. He is an inventor of ship shapes, a lobbyist for ship efficiency. An indication of his place in oceanography is the name of Woods Hole's submersible, *Alvin*, and its tender of many years, *Lulu* (Vine's mother's name). What irritates Vine is that ships are too often considered acts of God. He wants them analyzed the way oceanographic instruments are analyzed — in terms of cost effectiveness, data rates, and performance in such stressful, high-energy areas as the Gulf Stream and the high latitudes of the North Atlantic in January, which are nurseries for many a major oceanographic and meteorological phenomenon. In recent years a number of coastal research vessels have been launched. They have generally been smaller than the bluewater boats, not much more than a hundred feet. They cost less and require a smaller crew, but they also carry fewer scientists, often at greater risk. Waves over the continental shelves can be every bit as big as waves in the open ocean, and when they are, the smaller vessels either can't do much work at sea or have to stay in port. These are the craft that can scare you to death long before they drown you.

Vine saw innovation in the research fleet lagging several years ago and started complaining in print that the slowdown could do harm to what he called the triad of oceanography: ships, buoys and instruments, and satellites. "Without good ships," he told me, "we're not going to have good buoys, and without good buoys we're not going to be able to get the ground truth the satellite people need to make their information sparkle."

"The cost of a ship," Vine once wrote, "is small compared to the cost of operation for ten years and smaller still compared to the cost of the programs that may succeed or fail depending on the suitability of the ship." He thought it made sense for ship designers to improve operation capability by one sea state every three years. Right now, many ships can work fairly well in sea state four, although as things get rougher it becomes harder to recover instruments than to deploy them.

Sea states are convenient parameters for judging wind and water. At force five on the Beaufort scale, fairly common in winter in higher latitudes, winds are fresh breezes, seventeen to twenty-one knots. Waves build to six feet or so, with whitecaps and some spray, and life aboard takes on a decided spring. Force six is a strong breeze with ten-foot waves. Force seven, which

Vine would have his designers dealing with effectively at the end of a decade, is a near gale, with winds around thirty knots and fourteen-foot waves piling on each other, their flanks striped with foam. We were enjoying force seven and a bit more during that Georges Bank cruise. We got wet working in that. But seven is a long way from twelve. At that end of the Beaufort scale, winds blow so hard they plow the sea surface and turn the air above it white with water. This is a full hurricane, with seas of forty-five feet. Research vessels run from them, yet hurricanes, relatively rare as they are, are enormously important engines of heat and water transport. In one day they can effect more change, impart more momentum to the ocean, than can a year of light airs.

Cutting ship motion is a prime objective in research vessel design. Existing ships can be fitted with roll reducers, and new ones can be fitted with sails to add stability and cut fuel costs. *Sail* isn't really the proper word here. Designers are talking about vertical rotors spun by the wind that power propellers and other gear; about devices shaped like airplane wings on end. A veteran oceanographer who had been studying what might be done told me that cambered airfoils could drive a research vessel just fine on a proper reach. "The theory of flight's been solved," he said. "Now all we have to do is transfer it to the naval architects."

The oil industry has demonstrated the extraordinary stability of the semisubmerged hull. I have been on a couple of rigs drilling the outer continental shelf, inshore of the Gulf Stream. On one of them, a winter low came charging up the coast and presented us with a classic force twelve. Granted, the rig was huge. Granted, she broke two of her eight anchor wires. But she maintained her dignity in that madness, moving slowly, even gracefully on huge pontoons eighty feet or so below the torn surface. The same approach has been tested in small hulls with similar results, and it has become clear that a semisubmerged research ship two hundred feet in length can easily be as stable — and provide about as much working space — as the conventional hull twice that long.

Fully submersible research craft, of course, have been around for decades, and to them we owe a great deal of our detailed knowledge of ocean bottoms. *Alvin*, for example, explored the Mid-Atlantic Ridge, the spreading center between plates of oceanic crust, and later made some of the earliest chemical and biological measurements of warm-water vents in the sea floor. One sub actually rode the Gulf Stream, settled in it like some giant plankter and carried six people down the current. They saw salps in profusion, were attacked by a broadbill swordfish, and were visited by bluefin tuna and by the numberless organisms of the deep-scattering layer, the carpet of life that rises toward the surface by night and sinks by day. They had the same trouble staying in the Stream that any drifter does: after twelve days the current spat them out into an eddy, and they had to reinsert themselves. The sub was well named: the *Ben Franklin*.

Although the sea has served the imagination well, as any reader of Jules Verne knows, this is not the best of times to turn thought into new hulls and gear. The sixties and the early seventies were the top of the wave for oceanography. The science became multinational, multi-institutional, multidisciplinary, even multiship, as scientists focused on oceanic processes at the larger scales — general circulation, pollution effects, tectonic movements of the sea bottom. From 1958 to 1978 the number of U.S. institutions granting degrees in oceanography went from five to fifty, while the number of degree-earners grew fourfold, to about three thousand. Today, the number of ships in the national research fleet is down, and so is funding for them in terms of constant dollars. Add to this the natural conservatism and caution of the maritime mind, and odds lengthen against many imminent solutions to the ancient problems of going to sea.

Michael Faraday, the great nineteenth-century British scientist, once wrote that he "was never able to make a fact my own without seeing it." And that is precisely the central problem of oceanography: how to see the properties, phenomena, and functions of the seventy percent solution, that unseeable medium. The answer appears to lie in surrogates: the ship, yes, but more important, the contrivances of metal and plastic and glass lowered from the ship to do the seeing.

The maker of oceanographic instruments must have the skills of the finest watchmaker and the most imaginative armorer. His product must report what it sees with fidelities often approaching a fraction of a degree, a thousandth of a second. If it must float at the surface, an instrument must withstand forces operating in both air and water. Waves raise hell with it, causing it to flex up to a million or more times a month. If deep work is to be done, a device can encounter pressures read in hundreds of atmospheres. Corrosion is always a given; seawater will work its way in through a joint that looks vice-tight, and one drop is all it takes to scramble the signals. Marine creatures adopt instruments left too long in their habitat, fouling sensors and receptors. They did that to the tripods we were trying to recover out on Georges Bank; all one morning a technician hunched over a console trying to reach through the living mat on the tripod to a float release. "C'mon, machine," he crooned. "C'mon." It never did.

Some fish like to bite propylene line, so wire is used down to about eighteen hundred meters in many areas of the ocean. From time to time a billfish will develop unfriendly feelings toward a large object trundling through its domain — say, an *Alvin* or a *Ben Franklin* — and charge like a Miura bull. And more than once an enterprising ship captain has sidled up to a piece of equipment floating in the middle of an empty sea and liberated it. Nobody can prevent theft, but today's instrument designers are becoming quite good at minimizing malfunctions of other sorts — considering.

This expertise is fairly recent. It was only a little less than twenty-five years ago that the Oceanographic's Henry Stommel did some grousing about the state of his art. He was out in the tricky Somali Current, off the eastern coast of Africa near the Gulf of Aden, when he wrote:

> It is rather quixotic to try to get the measure of so large a phenomenon armed only with a 12-knot vessel and some reversing thermometers. . . . Nothing makes one more convinced of the inadequacy of present-day observing techniques than the tedious experience of garnering a slender harvest of thermometer readings and water samples from a rather unpleasant little ship at sea. A few good and determined engineers could revolutionize this backward field.

Sound is one thing that does penetrate the sea with ease. Temperature and pressure combine to lower the speed of sound down to a thousand or fifteen hundred meters — the general location of the Sound Fixing and Ranging (SOFAR) channel. Below that, increasing pressure serves to raise sound speed. Sound rays bend down into the channel from above and up from below, and tend to stay in it for considerable distances. Just a few watts' worth of sound can be picked up a couple of thousand kilometers away. We are newcomers at the game. A cetologist at Woods Hole told me once that among the noises to be heard down there are the voices of whales. "You know," he said, "one off the Bahamas pipes up with, 'I'm Moby, I'm Moby,' and another off Africa says, 'I'm Dick, I'm Dick.' "

It was John Swallow's sound-tracked floats that helped find Stommel's undercurrent, thereby tearing a hole in the conventional notion of abyssal passivity. No longer was it possible to get a handle on deep circulation by sending out a few ships and averaging the readings: that amount of motion had to be studied by new instruments. As so often happens in science, observations were forcing theoretical shifts, which in turn were forcing new observations. (The very rare exception to this pattern was Stommel's positing of the undercurrent *before* it was found.)

Oddly enough, though the floats discovered the strength of benthic currents, oceanographers did not at first use them much in following up on the discovery. Floats drift with the water. A more comprehensive system, it was thought, would be a pattern of stationary instruments measuring the flow past their moorings. The call went out to develop current meters that could record those flows, accurately, over many months, and by the early seventies, such meters were available. One of the most promising was a vector-averaging instrument that smoothes out the twists and turns of the water and records on tape what moves where and when on a horizontal plane, the dimension the ocean prefers for most of its traveling.

That meter and others are nothing without their moorings: kilometers of wire and line, with instrumentation at precisely planned intervals, interspersed with flotation clusters — usually "hardhats," plastic covers bolted over special hollow balls strong enough to take the pressure without imploding. Some moorings run from anchor to surface float, which is often a big, doughnut-shaped affair with a tripod on top equipped with a strobe light, radio antennas, perhaps an anemometer for the wind. The mooring often preferred by the buoy group at Woods Hole is subsurface, the top of the string coming to within a couple of hundred meters of the surface.

Launching any deep-ocean mooring requires the finest ship

handling and positioning. The first comes from years of practice and the development of cycloid propellers, bowthrusters, and other components of modern propulsion. The second depends increasingly on a satellite navigation system that can put the chief scientist within a few meters of where he wants to be. Ship speed is called for in tenths of knots. The trick is to stream the mooring aft without a kink or snarl and come up on the deployment point without too much towing, particularly in rough seas. When that point is reached, the mooring's anchor is wrestled over the stern and begins its plummet through the miles of water.

The floats and sensors follow, sometimes in spectacular fashion. I remember watching one subsurface mooring being paid out. A great orange ball had been attached at its top. After deployment the deck boss in charge of the array had the ship bear off. In the excitement he delayed giving a new heading, and the ship came around in a curve. No danger was involved, but we saw a wonderful sight — the water-sheave effect, in which the sinking anchor draws the rest of the array to the spot where it splashed in and then down, following its trajectory. We looked to starboard and saw the great bright float rush over the sea toward us like some mutant whale and then disappear. I thought of the great array settling, coming to rest in its appointed place, nodding in the abyss.

Getting all that expensive equipment back again was fairly chancy a couple of decades ago, before the appearance of acoustic releases that respond to a signal by firing an explosive bolt or in some other way uncoupling the array from the anchor and allowing the flotation gear to bear it up to the surface. The Woods Hole buoy group now recovers more than ninety percent of what it sets, but machinery still goes on strike, as it did that time on Georges Bank, and the ship has to go hunting or leave hundreds of thousands of dollars' worth of equipment, and the priceless data it is recording, to the sea.

Moored arrays are difficult to deploy in currents like the Stream, and what Lieutenant Pillsbury did a hundred years ago turns out to have been more a *tour de force* than the beginning of promising experimentation. However, new designs have made it possible to anchor moorings under the current and to place sensors within a few hundred meters of the surface, in a zone where the water moves at a respectable two to three knots. From there, instruments like the Doppler sonar, which measures shifts in back-scattered sound, can get accurate readings of the high-speed core rushing by overhead. Out beyond Hatteras, where the Stream begins its meandering snake dance, moored arrays are still useful for measuring the fields in which the current moves, though they will miss the current itself when it shifts away from their positions.

Shipboard profilers can be used in swift water, as Fritz Fuglister so aptly demonstrated. New models can yield a temperature-with-depth profile over hundreds of meters. The CTD, for conductivity-temperature-depth profiler, gives a more intense data picture, often reinforced with water samples drawn by special containers set in a rosette around the instrument. These are the grandfathers in a full house of instruments designed for high-energy environments.

World War II gave modern oceanography its first technological boost, particularly in the use of underwater sound. The space age has given it the second. New metals and plastics are going into instrument housings and new electronics into their innards. Gone are the old days of unreliability, when wags suggested that the ideal oceanographic instrument should consist of less than one vacuum tube. Miniaturization is now old hat, and ocean engineers are learning how to produce reliable data over long time periods using a fraction of the power output once required. Some of the work is being done at European centers, which have produced such elegant devices as the batfish and its children — heavily instrumented, towed structures that can be flown up and down in the water column, yielding a continuous picture of ocean performance over long distances. In the United States any number of research centers, from the University of Miami to Scripps Institution of Oceanography, are in the game. Woods Hole is too, of course, and the University of Rhode Island. Techniques proliferate. New rotors and lasers measure vertical mixing. New remote-controlled vehicles carry new cameras into ocean-floor fissures unexaminable in any other way, or down the ravaged passageways of the *Titanic*. Tubes and spheres free-fall to the bottom and bounce up again, measuring, measuring.

In its extensive Gulf Stream work, the University of Rhode Island often utilizes a fairly simple instrument to study oceanic complexities. The inverted echo sounder is little more than a transceiver on the bottom. It sends out a sound signal which bounces off the surface and returns. Changes in the average temperature of the water passing over the instrument affect the speed of the sound signal, reflected in the time of its passage up and back. The instrument can judge movements of thermoclines and fronts, where waters of sharply contrasting temperatures abut.

Randy Watts is the man in charge of the instrument at Rhode Island. He and his group have discovered that if they can keep track of the 12-degree-centigrade isotherm, they can learn a great deal about how the Stream meanders. This isotherm, the imaginary line connecting points at which seawater measures 12 degrees, is the telltale for the center of the Stream's thermal front. Watts and his group have put out arrays of echo sounders combined with current meters east of Hatteras and left them there for months to record temperature changes. Borrowing a technique called objective analysis from the meteorologists, they were able to establish the depth of the 12-degree isotherm sufficiently well to map it on a daily basis for three years, starting in 1982. This in turn enabled them to watch the Stream as it passes what they call the Inlet, where it pours off the Blake Plateau near Hatteras and begins its penetration of the open ocean.

Watts likens what they have seen to what meteorologists study in the atmospheric jet streams. The Gulf Stream is about ten times smaller than these high torrents, and its variations occur ten times more slowly. But both systems exhibit narrow bands of rapid movement. When speeds diminish in those bands, meanders grow and move downstream. The meanders shed eddies and other phenomena that ultimately have an effect far from their breeding grounds. Meteorologists are sufficiently comfortable with their eddies to forecast storm directions and intensities three or four days in advance. Watts and a growing number of oceanographers feel that they are now at a point in their understanding where they can begin to do the same for the Gulf Stream and, ultimately, the world ocean. They are where the forecasters were twenty or thirty years ago. Given the medium they work in, they are proud to be there. A three-day prediction for the atmosphere is the equivalent of a twenty- or thirty-day prediction for the far denser and more ponderous ocean, and that time scale no longer seems unattainable. The Office of Naval Research, a prime funding source for ocean physics, has launched a five-year program called SYNOP (for Synoptic Ocean Prediction) to help reach that goal. Its target area is the Stream.

Watts works closely with a man Henry Stommel might well have had in mind when he put out his call for some top-notch engineers. H. Thomas Rossby is an engineer *and* a physical oceanographer. He grew up listening to discussions of how fluids move. His father, the famous Swedish dynamicist Carl-Gustav Rossby, devoted a great deal of time to the workings of atmospheric jet streams and some to the oceanic variety, leaving a trail of Rossby numbers and Rossby waves and mathematical analyses of flow.

Working with some of the best instrument builders in the country, Tom Rossby has fashioned devices that behave much like the meteorological instruments his father worked with and in some important ways, better. Meteorologists have problems with their balloons. They can't fly them at low altitudes because of danger to aircraft, and icing problems make it difficult to operate them at other levels. Flying balloons in the sea has its advantages, Rossby told me when I first went down to see him at the university's Graduate School of Oceanography, overlooking Narragansett Bay. The ocean is a comparatively stable environment, and, he added, "There's nobody in the way." He laughed, and the look on his face put me in mind of an arctic fox. "Our float technology," he said, "is way ahead of any similar balloon technology in the atmosphere."

Rossby worked on some of the successors to John Swallow's great tubes, like those used to track the slow swirls of large eddies in the ocean interior. About ten years ago he began to concentrate on float measurements of the Gulf Stream in order to learn more about its velocity structure and movements along its north-south, east-west, and up-down axes. The proxy for these studies was a free-falling device that talks to transponders on the bottom as it sinks through the water column and then bobs back to the surface. Signals sent and received are recorded so that the instrument's precise travel paths can be plotted. Rossby called the device Pegasus. In the early eighties he crisscrossed the Stream with it, concentrating on an area favored by Randy Watts and other URI investigators, about two hundred kilometers east of Hatteras.

Pegasus suggested to Rossby that velocity structure remains remarkably constant as the Stream bulls its way through the sea. On the northern, or inshore, side, the current is strong and shallow. Most of the transport occurs on the southern side, where the current is slower, broader, and deeper, extending below two thousand meters. Water is added to the Stream from both sides, and that may help account for the sharpness of its delineations. These injections, which occur at all depths but appear to be at their strongest far below the surface, may also be the source of the Stream's deep transport, since the Straits of Florida are much too shallow to allow deep water from the Caribbean or Gulf of Mexico to pass through the fire hose.

Floats moving at two thousand meters under the Stream had shown pretty conclusively that the current bottomed there. They went off every which way, while the "river" above them headed eastward. But the floats, and some elegant work on water mass analysis that tended to corroborate their data, couldn't give much of an idea of how deeper currents moved with respect to shallower ones. That kind of work called for a new drifter, a jewel of a thing a little more than five feet long, a tube of Pyrex glass housing a vision of miniaturized electronics, all copper and color. Rossby and his collaborators called it RAFOS — SOFAR backward, a name as precise as the instrument. SOFAR floats broadcast their positions to an array of listening stations; RAFOS drifters record incoming signals from moored sound sources that broadcast every eight hours. They record that information, along with data from temperature and pressure sensors, and at the end of their tour — usually forty-five days — they release ballast, surface, telemeter their findings to an American satellite, which passes them on through a French communications system, and voilà!

Given the immensity and the complexity of what they're dealing with, most oceanographers have to remind themselves that what their data show them is a picture limited by luck and the biases of their instruments. "We know what [our] gear will catch," one of them said a decade ago, but "we do not know what it will not catch." Rossby and others familiar with the SOFAR floats knew that although the neutrally buoyant tubes would follow paths of equal pressure, sticking pretty much to the same depth, a given water parcel may not do that. Water tends to follow density surfaces up and down. Density surfaces, at least in the upper ocean, often cross pressure surfaces. Later models of RAFOS thus incorporate a simple spring-loaded piston, which gives the float the ability to expand or contract, to mimic the compressibility of the surrounding fluid. As its home water rises or falls with a density surface, RAFOS goes with the flow.

The floats can be launched from ships of opportunity, an enormous advantage, given costs and competition for ship time aboard research vessels. A fishing boat did the first work, in 1984, but the Stream was too rough with it, and Rossby's group shifted to a small freighter on the Norfolk-Bermuda run. They gave the skipper an advanced version of Fuglister's profiler called the expendable bathythermograph, or XBT, and asked him to launch his floats when, say, the 15-degree isotherm intersected four hundred meters, and the game was on again.

The floats, deployed at the Inlet, near Hatteras, sank to the Stream's thermocline and moved off, a few on utterly unlikely tracks, some circling in the thrall of an eddy. But many stayed with the main current, and Rossby began to see patterns. One RAFOS, tracked for a month of meandering, followed water at 10 degrees centigrade as it rose and fell between four hundred and eight hundred meters. Other tracks, added on, showed that the floats all rose in the water column as they rounded a meander bulging toward the coast and fell in a meander extending toward the Sargasso. The wellings sometimes were strong enough on both sides of the current to eject water from the Stream. At other times, they were barely noticeable.

Rossby wants to know why that should be. He also wants to train his floats to osculate — not to kiss, but to bounce up and down between two adjacent density surfaces, measuring the changes in the vertical distance between them. If he's successful, he will have added a new tool for looking at the stretching term, an important function in fluid dynamics dealing with the spin of water relative to its position on the globe. If you know the stretching term to closer approximation, you have moved well forward in understanding the dynamics of the Stream.

Sound is the *primum mobile* of another, more ambitious idea among the instrumentalists of oceanography. It is called ocean acoustic tomography, and though it is designed for use in ocean basins, it is also being tested in the Stream. It is much like medical tomography, the computer-augmented, X-ray sections of the CAT scan, which reveal what is going on in the brain or other delicate organ without invading it or seriously affecting it. Seismologists use a version of tomography in trying to study the mantle and core of the earth by recording the travel times of earthquake tremors through the planet. Iteration is the key. They start with a generally accepted idea of the way things are thousands of miles beneath the ground. As they build up their seismic records, they continuously amend their models. Eventually, what they see in their models should match what they see on their instruments. The exercise is known as inversion, and oceanographers believe they can practice it to considerable advantage in their medium.

Carl Wunsch and Walter Munk began thinking seriously about the possibilities of ocean acoustic tomography in the seventies. Wunsch is a physical oceanographer at MIT, in Massachusetts, and has close ties with Woods Hole and other marine science centers. Walter Munk, one of the most respected of senior oceanographers, is based at Scripps, in California. The two men began talking tomography to sound specialists here and there, and gradually talk began turning into casings and electronics and testing. A test in which signals went only from instrument A to instrument B was carried out in 1981, off Bermuda. Two years later the tomography team set up a reciprocal array south of the Stream, one in which sound shot back and forth among several instruments, forming a sentient fabric of acoustic rays.

Munk and Scripps have been running tests in the Pacific, experimenting with low-frequency sound that can reach far out — the moorings are a thousand kilometers or so apart. A lot of processing is needed to produce data of usable resolution. Wunsch and Woods Hole are working on a smaller, less expensive, but more limited instrument. Components come from groups at MIT, the Oceanographic, and France. In 1987 some of the design and much of the assembling and testing for sea trials was going on in a tiny wooden building outside Woods Hole, headquarters for the Webb Research Corporation. Douglas Webb, president, is a Canadian who has built a name as an instrumentalist. He worked in Europe for a while, then spent twenty years at the Oceanographic developing, testing, and building floats and moored devices. Now he employs himself, a talented French engineer named Pierre Tillier, and a few others to work essentially on just two devices. The first is a bobber, a deep drifter that surfaces periodically to let a satellite know what it has been doing and then sinks to do some more. That is Doug's project. The second, run by Tillier, is tomography.

The speed of sound in the sea, around fifteen hundred meters per second on average, changes with temperature, pressure, salinity. So if you can devise a way of sending sound waves to a receiver a sufficient way off and then analyzing that sound for perturbations in the time it takes each pulse to travel its path, you should be able to extract information about the properties of the seawater along those paths. You should also be able to detect actual water movements: a current flowing in concert with a sound pulse will help it along just a little; one flowing against it will brake its speed.

The heart of tomography is a clock. An atomic clock would be best, but that draws too much power for a device that must remain submerged on its mooring for perhaps a year or more. Instead, Tillier and others are working on a more conventionally powered clock that "drifts" no more than a few milliseconds a *year*. Everything depends on time reduced almost to a point. Even the mooring is timed. Acoustic beacons around its anchor are queried before a pulse is transmitted so that a navigating device in the electronics pod can figure its position. The mooring is designed to stand tall, with flotation devices exerting more than a ton of upward pull on the cable. Even so, moorings do nod, and a lateral movement of even a meter and a half will add or subtract up to a millisecond of pulse travel time.

Much of the timing device Tillier uses for equilibration of instruments is housed in a refrigerating unit that takes up a fair amount of the lab. The rest is workbench, electronic gear, and computers. The queer stunted oaks of the Cape catch the wind outside his window. Inside, all is quiet, miniaturized. Tillier described the electronic package of the transceiver in his precisely accented English: "You can see there is a transmitter which is generating a code and then matching the amplified signal to the transducer." The transducer is mounted in a tube outside the main housing and uses a diaphragm activated by ceramic disks to create the requisite pulses. "There is here a recorder. There is here the clock. There is here the receiver." Downstairs, in what must once have been a garage, are a few long gray barrels that look vaguely like rocket launchers. The electronic pod slides into them, the ends are sealed, the transducers fitted on, and the instruments start on their way to sea. The ones I saw were going to France for tests. The next batch would be for Wunsch. All the tubes would be deployed around the Gulf Stream later on as part of the SYNOP experiment. If they work well, they will be man's first seaborne proxies to yield basinwide, synoptic results, the first to take a picture of an ocean from the inside.

It will take time for confidence in tomography to build, if it does build. Many new instruments present a picture of ocean functions that just doesn't fit with the accepted image. The current meter was suspect for a time because its readings appeared so different from those derived from geostrophic calculations, but after a while the fit got closer, and the meter joined the fraternity. Ocean tomography gives something of a smeared picture. That will improve. But when it does, how will you calibrate what you get against accepted standards when no instrument exists that can deliver a daily interpretation of how things are in a patch of ocean a thousand kilometers across? "How do you get to believe in the picture?" Webb asked. There are other problems, as there would be in such a technical departure. A full-basin array of, say, ten or more moorings, each with its transceiver standing in the deeps (most a bit shy of one thousand meters) will cost a lot — well over $150,000 per mooring — and generate Himalayan amounts of data. Analyzing it will require very large teams at very large institutions. So will maintaining and monitoring it.

Right now, the data are retrieved the way most recorded data are retrieved: up over the side of a ship. But that means months, perhaps more than a year will have gone by with no word from the transceivers. Are they working? What have they discovered? The sheer amount of data militates against telemetering, unless some processing could go on inside the transceivers of the future. If that were done, the standard procedure would be to broadcast the results to a satellite via a surface link. But because the ocean surface is often such a tough place to be, perhaps some way can be found to transmit data acoustically, to use the sound that investigates the sea to carry the results of its investigation. Tom Rossby is working on that idea for his RAFOS floats.

All in all, Webb believes he has never seen such a window for advancement in oceanographic instrumentation. And tomography holds a particular promise. We want to observe the oceans in both spatial and temporal dimensions, he says. A ship is great in studying the vertical and it can move around some in the horizontal. But it is bad in time terms; it can't hang around long enough to study variability. Current meters and temperature-pressure sensors do well in the vertical and can stay in one place for many months. But they can't cover too much ocean without prohibitive cost for multiple arrays. "Tomography," Webb says, "is the only one that is fairly good in all four dimensions."

I didn't pay enough attention to Seasat when I had the chance. It was 1978, I had just fallen in love, and I can tell you that sheep's eyes and a long sea voyage are not compatible. When I wasn't trying to reach the object of my longing by radio — a process made almost impossible by our location several hundred lonely miles west of Scotland, and by the occasional and always irritating yammer caused, I was told, by interference from long-range Soviet radar — I moped around so that I had trouble concentrating on the oceanography at hand.

JASIN was the name of the project, for Joint Air-Sea Interaction Experiment. Research vessels from the United States, Britain, West Germany, the Netherlands, and the Soviet Union bobbed and curtsied to one another in a patch of sea around Rockall Trough. The experiment was designed to learn more about how heat flows from air to ocean and back again, how the winds impart their momentum to the sea. JASIN planners had deliberately picked one of the stormiest regions they could find (a magazine article on the bulletin board of my ship claimed that in theory two areas in the Northern Hemisphere were capable of producing waves more than two hundred feet high: the Gulf of Alaska and the water not far west of us). The odds at Rockall Trough were fifty-fifty that during two months in the late summer up to seven windy lows would come blowing through. Only a couple did.

I don't think I will ever again see the concentration of instruments that went over the side and up in the air at Rockall: miles of moorings, scores of towed and free-drifting devices, meteorological blimplets and radiosondes. Aircraft passed overhead, and way above, eight hundred kilometers over us, Seasat flew, the first and as of this writing the only satellite dedicated entirely to oceanography. She was a marvel, stitching her orbits, one every hour or so, into something truly approaching simultaneity. Her instruments recorded and transmitted millions upon millions of data bits that would be processed into pictures of the Gulf Stream, the Kuroshio, temperature anomalies in the North Pacific that might affect North American weather, ice conditions near the poles, and the wind at Rockall Trough. Some damfool short circuit took her out long before her time, but during her hundred days she gathered more information on sea-surface phenomena than had ever been gathered by one vehicle before, more information on wind speed and direction, to name just one field, than had been collected over the previous hundred years aboard ships.

Four years before, on a research cruise poking around the sills of the Caribbean, I had seen a satellite come right over the masthead, heading north, a poky meteor. Perhaps on a rare night without clouds, perhaps with binoculars, I could have spotted Seasat's insignificant gleam. But as I say, I was preoccupied. I never looked up.

The idea of looking down at the ocean goes back a long time. Lookouts stood at crosstrees for centuries, watching for wind and whales and the signatures of currents or breakers. Oceanographers have been using airplanes since they could be trusted to fly out to sea and get back. Henry Stommel and some colleagues used one to take the temperature of the Stream in the mid-fifties. Their instrument was a radiometer. Fifteen years before, a scientist had pointed a similar device in the general direction of the Florida Current from a hotel window in Miami Beach; he got the signatures of wakes, streaks of cooler water evidently stirred up by ships' propellers.

Then meteorologists began using radiometers to get pictures of cloud formation from space. The first satellite so equipped went up in 1960, and the weathermen soon found that when no clouds were around they were picking up emissions from the surface of the sea. The public paid attention to the glorious photographs astronauts were taking from hundreds of miles up. We saw, quite suddenly, that we were riders on the earth. But scientists, who wanted not snapshots but synoptics, images produced from a stream of data received from orbit after orbit, began to work hard on the radiometer. They expanded the range of its wavelength from visible light to infrared to microwaves. So did broadcasters, the military, others, with the result that the planet is now aglow with transmissions that can interfere with some satellite sensing. An enterprising alien with a radio telescope should be able to spot us from ten or twenty light years off.

Radiometry has produced some mythical sights, none more so to my eye than Gulf Stream imagery. When computers became deft enough to turn numbers into images, we saw a Chinese dragon off our Atlantic coast, flaming at the warm core of the current, cooling through the colors at the margins. At first, instruments dealt only in relatives: this water is colder than that. Now, advanced units can measure actual temperature to less than 1 degree centigrade, and the error is shrinking. Like almost all oceanographic instruments, radiometers measure indirectly. It is sometimes hard to tell what precisely they are looking at and how what they sense relates to the actual temperature. The sea surface is one of the most complex structures in nature, and radiometers cannot pierce it to see below. They play with the top micrometer or two, whereas ships measuring the surface probe several meters down. Things like that make a difference. What the radiometer sees may be useful for experiments in air-sea interaction — the stuff of JASIN — but not for other, equally important analysis.

Clouds have been a problem, too, thin high ones or those just over the surface too small to be spotted. They have caused low readings. Bigger and thicker varieties have blanked out the picture entirely. But passive sensors, receptors of electromagnetic waves, have come on to the point where they are contributing to global mapping of actual sea-surface temperature. So have computer systems for processing data. One of the most popular was developed at the University of Miami. Otis Brown, part of the group that did that work, has used the system to create a library of color slides and film loops that thermally trace surface currents, including major western boundary currents — the Brazil, the Agulhas, the Somali. Brown showed me how every year the Brazil Current snakes down off the Argentine coast south of its normal position, gets cut off, and retreats. I saw El Niño in bands of color, the disastrous warming of the eastern Pacific. But the phenomenon was more than regional. While the waters off Peru were heating — and fish were dying — the equatorial Atlantic was cooling. "That's the signal," said Brown.

Peter Cornillon works with the University of Rhode Island's oceanographic remote sensing unit, which in turn works with software from Otis Brown's group in Miami. Cornillon has used data from modern infrared sensors (carrying such optimistic names as Advanced Very High Resolution Radiometer) to have some very serious fun. When hurricane Gloria came whipping up the eastern seaboard of the United States, Cornillon and his group examined how cool water welled up in its footprint. They mapped outbreaks of warm water from the Stream into the Sargasso Sea. And they have been using years of satellite data to create a diary of Stream movements, especially those off New England and the Maritimes.

The diary has been converted into television images. A yellow serpent on the screen representing the sharp northern edge of the current wanders with the weeks. Meanders appear and grow, and soon eddies will be part of the show. Cornillon suspects that the rings have an influence on how and when the Stream meanders. His data, checked against numbers from instruments operated by Randy Watts and Tom Rossby and others covering roughly the same area, provide a new way to study the Stream. He can work almost in real time, with information just a day or two old, and he can refer back to historic Stream positions.

Sometimes Cornillon's screen shows moves that are hard to dispute, like the pronounced northward shift he found in the Stream's envelope, starting off Hatteras and, over four years, extending several hundred miles downstream. Sometimes his findings even go against those made with in situ instruments. For years oceanographers have thought that the New England Seamounts exacerbate the Stream's lateral shifts. Phil Richardson, the Franklin fan, has reported a lot of variability over and downstream from the seamounts. But Cornillon's data say no, there is no surface signal of such goings-on in that area. If there is any effect, it may be upstream rather than downstream. The two sets of findings aren't unalterably opposed, Cornillon argues. He used figures from more than two years of satellite passes over the seamounts. Statistically, the seamounts may not influence meandering of the Stream to any large degree, but in a particular instance, a particular time, they may.

The University of Rhode Island has tried marketing its remote-sensing skills by working with Shell Oil when that company decided to drill test wells in six thousand feet of water far out to sea in Baltimore Canyon. The project was right at the edge of the technology of the time. A drill string of that length is a fragile thing, vulnerable to fairly sudden changes in currents. With one exception, URI was able to warn the rig when eddies from the Stream were bearing down on it. The contract called for only intermittent observation, but a tracking buoy was deployed in the wrong ring, and the rig was hit by a powerful eddy at just the wrong time. Shell then called for continuous monitoring. The oil rig found nothing worth producing but suffered no more ring damage.

Cornillon's team also mailed charts of sea-surface temperatures to some New England fishermen, who said they were helpful though spotty (again, the cloud problem) and a little too outdated for their purposes. Meanwhile, a federal oceanographer named Jenifer Clark was developing her own "Gulf Stream Daily," a chart of the current system based on infrared data. Clark's analysis, offered on a subscription basis by the National Oceanic and Atmospheric Administration, is a product of satellite and sagacity. She expects to get analytical help from a computer program some day, but right now she is the analyst. Twice a day she gets her satellite images, interprets them, and charts them. I have a copy of a "Daily" on the wall in front of me, the one for June 27, 1986. I see temperature readings all over the place. A heavy line marks the north wall, and there are other symbols for warm and cold eddies and for current flow. All the major submarine canyons are marked, from Baltimore to Poor Man's to Corsair.

Swordfishermen take Clark's publication because they need to know about the rings and the other fronts. So do longliners and sports fishermen after other pelagics. Lobster fishermen want to know when a warm-core ring is coming, since one powerful enough to scour the bottom can make off with their gear (a theft they once blamed on Russian boats that used to harvest American shelves and slopes like combines in a grain field). Fisheries managers are trying to decide whether warm-core rings really do swirl into spawning areas and carry fish eggs and larvae off to perdition. These managers had access to imaging from the spectacular Coast Zone Color Scanner, an orbital sensor that, while it functioned, painted abstractions of chlorophyll concentrations sophisticated enough to indicate zones of high and low productivity—relative concentrations of plankton. Now they read the "Gulf Stream Daily."

Ocean dumping is still going on, and the Environmental Protection Agency and the bargemen need to keep track of rings. There is a site a hundred miles off Cape May, New Jersey, where if you dump when a warm-core ring is swinging through, the clockwise current will carry your wastes, some of which are quite nasty, toward coastal waters. You want to wait till it goes by. The same sort of thing, on a larger scale, applies to plans for incinerating wastes at sea. You will be obliged to know what both wind and water are doing before you light the match.

As with *Exxon Wilmington*, big ships can save big money in time and fuel by keeping track of the work of Clark and other oracles. So can sailing vessels. Ocean racers bound for Bermuda once got briefings from people at Woods Hole, who used to fly over the Stream a day or two before the starting gun went off. Now the sailors call Jenny Clark.

Seasat carried a microwave radiometer. She also carried three other instruments that in various combinations are likely to give oceanography a considerable lift in its desire to become, like meteorology, a truly global science. One was a scatterometer, one a synthetic-aperture radar, and one an altimeter.

The wind begins. It teases the sea into a roughness, raising wavelets on the skin. The wavelets grow to waves, and the waves run before the wind. Slabs of sea move before the wind and, in the North Atlantic, curl to the right. At some point over the fetch, the sea breaks from under the wind and enters a larger regime, that of the general circulation of the upper ocean.

Scatterometers can make sense of the wind and don't have to wait for a clear day to do so, though rain on the sea surface can confuse them. They measure roughness, the wavelets raised by the wind that scatter the satellite's beam. There is still some debate as to the precise relationship between wavelet and wind speed, and there is still difficulty in validating scatterometer readings, since they are averages. But precision is improving. The Seasat scatterometer measured wind speed to within several knots, but it was ambiguous about direction. It could say the wind was moving on a north-south line, but it couldn't say which way. Nonetheless, it demonstrated the inaccuracy of most of the old wind charts, taken as they were from readings by eye or from poorly placed instruments and entered in thousands of logs by thousands of exhausted and distracted mates. The newer models produce data that are much less ambiguous and that, properly processed, can show what one client says are "new and intriguing

swirls, bands, rolls, and vortices that we only suspected to exist before."

A synthetic-aperture radar is a radar that cheats. Standard radar antennas are too short to give good resolution, so electronic processing of the signals has been developed to simulate antennas that would be a mile or so long if they existed. With SAR, oceanographers can develop indirect evidence of undulations along thermoclines and other such subsea surfaces. SAR also works well over the Stream.

So do altimeters. These radars look straight down, measuring the height of waves, of the entire sea within the relatively narrow scope of their sensors. The Seasat altimeter was accurate to within about fifty centimeters. That was good, but not good enough for the kind of work oceanographers like Carl Wunsch have in mind. In cooperation with the National Aeronautics and Space Administration, which has developed and launched most of the civilian satellites and sensors operating for science, oceanographers have been working toward an altimeter that can read the sea in a way that will turn dynamics on its head.

This planet is not the marble we conjure up. It is full of lumps and bumps, thanks to variations in mass and therefore in gravity. Stop the earth, get off, and regard it from a distance. In this fantasy, if equipped with properly fantastic instruments, you will see that the surface of the sea, now motionless, undulates over a range of about 185 meters, a rough map of the rises and trenches that lie below. That is the geoid. It has not been properly plotted yet, but the rough shape is known. Now start things up. With your X-ray vision, you can see some small shifts occurring in sea-surface height. A meter or so is due to tides, a few centimeters to changes in atmospheric pressure, the rest to geostrophic surface currents.

An altimeter with sufficient sophistication could read the sea and give oceanographers the true height of sea surfaces around the globe. From that, they could use dynamical equations to get current velocities and directions. They would have no more need for theories like the level of no motion — where gravimetric studies have mapped the geoid. Where they haven't, sea height can't be precisely given. But even then, repeated satellite passes can yield changes in that height and therefore changes in the strength of currents. The geoid is well known in the northwestern Atlantic, and so the Stream could be monitored continuously for variability in surface velocity. More important, scientists could improve their understanding of how the Stream extends eastward, how it rejoins the sea.

Spacewalking gremlins killed off Seasat, and they continue to raise hob with other orbiters. Clouds and dirt and ionic mischief confound passive sensors, rain and wave troughs and other plagues do in active ones. For all the modern hardware, data have become a problem. Some radars put out information at the rate of tens of millions of numbers a second. Many instruments transmit material that requires elaborate interpretation to be of any use. Yet the chaos of the electronic age gives here and bends there, and things go on. The flood of data is drained away by computers so improved they begin to want for work. Then a new generation of instruments arrives, and the flood returns.

American space industry is still in a drift, and the viruses attacking the nation's confidence and currency can only complicate the difficulty. But it looks as if a few projects dear to oceanographic globalists will proceed. NASA is working with the French on a new altimeter program known as Topex-Poseidon. It is possible that a fine new scatterometer and supporting sensors, previously planned for a Navy satellite scrubbed at this writing, may go up in a Japanese bird. A European satellite, ERS-1, is scheduled to fly in the next year or so, and it will carry some useful sensors.

And the ships, the crews out on the fantails fighting the roll? Henry Stommel and many others fear that oceanographers of the future will compete more for computer time than for ship time. That may be so. But it is wonderfully ironic to me that when Seasat shut down and was just another piece of angular junk circling overhead, people had to go to sea to find out what she had meant to say. It probably is safe to predict that satellites and the new in situ instruments will reinforce the move away from the oceanographer as lone man and toward the oceanographer as organization man. But whatever he is, he will have to keep getting wet to get the truth about the sea.

8
Gulfcast

ONLY A FEW DECADES AGO, when computers were new and scarce, people building theories of how the ocean moves did a lot of their experimenting with a pencil. They worked mathematically, with simple constructs of flat-bottomed homogeneous seas in rectangular basins, the waters just ghosting along. Their oceans were slow. The great advantage of the slow ocean, then and now, is that it is linear. You apply a force here and you can follow its course right along to there. The flow is called laminar, something rare in a fluid. The bathtub provides a very rough illustration. If the tap is barely open, water will run down the drain without much fuss. You can see that it is ordered in its descent. As the tap opens, the flow down the drain soon turns chaotic, bubbling and burbling and whirling. Now, dear bather, you have nonlinear flow, turbulence. It is impossible to follow each filament and whorl. The amount of water emptying out of the tub can be measured without trouble, but not much else.

The linear constructs concerned themselves with the wind-driven portion of ocean circulation, and their simulated winds were weak enough to move the water in such a way that the observer could judge which current was responding to which component of the wind system. Peter Rhines, an ocean modeler at the University of Washington, once described a slow-ocean simulation of the North Atlantic:

> Here, in the model North Atlantic, the clockwise sense of the wind stress favors southward flow, and the water can return to the north only in the "shelter" of the western boundary. In a beautiful mathematical theory begun by Stommel, the Gulf Stream appears as a boundary layer, rather like the region very near an airplane wing where fluid velocity varies unusually quickly from streamline to streamline.

In a slow-ocean model, water moves slower than the speed of the wave patterns — between two and ten kilometers a day in midlatitudes. But real water is not often likely to poke along so. Its speed is such that waves can break and currents can turn nonlinear. Under real conditions, Rhines wrote, "the Gulf Stream, instead of smoothly losing its fluid to the open ocean as it moves northward, escapes the coast altogether. It is a mathematician's nightmare, with the boundary current escaping its boundary, and snaking around in the open ocean like a fire hose out of control."

The turbulence isn't completely three-dimensional, since the ocean moves largely in the horizontal plane. But it is pervasive enough to have limited oceanography to useful but distorted simulations of the oceans — that is, until the computer gained enough sophistication to begin measuring ocean reality in at least some of its multiple scales of time and space. Differential equations of nonlinear flow remain more or less unsolvable, but the computer can work with portions of equations and small bits of time, and the speed of its calculations makes it possible to simulate fast-ocean flows to a degree. Working with it, modelers have established themselves firmly in the middle of what Rhines called the scientific bucket brigade between those making observations in the sea and those making theory in their minds.

It doesn't seem right that one of the most advanced facilities for ocean modeling sits high on the Front Range in western Colorado. But that is where Rhines and others of his persuasion have often gone to feed prized notions into supercomputers. There, at the National Center for Atmospheric Research, scientists and artists of software build complex programs that seek to push accurate weather prediction forward in time, to better understand the cycles of climate, and to simulate the great air-sea engine of the earth.

William Holland is a resident at the center. His office in the handsome pueblo designed by I. M. Pei is so stacked with papers it looks like something out of Edgar Allan Poe. Holland is a physical oceanographer who is known for his ability to bring the ocean up on a screen, sometimes in four dimensions. One of his creations is a simulation of the Gulf Stream envelope, a loose weave of bright lines connecting equal values of this or that property. It can be rotated in its planes and set in motion to show the belly of the current undulating and eddies forming and reaching for the bottom far below like twisters touching down.

Holland has developed what is called an eddy-resolving model for the North Atlantic, and that is quite a feat. In the atmosphere, eddies — weather systems — can extend across thousands of kilometers and can be identified using fairly coarse grids, points at which variables like temperature and pressure are defined, coded, and given to a computer. Ocean storms move much more slowly than those in the atmosphere, in a fluid arranged in far different patterns of stratification. Time is not so much of the essence in tracking them, and they are also far smaller than their cousins above, running from a few tens to a few hundreds of kilometers across. They require a much finer grid than do atmospheric models. The finer the grid, the more data, physics, and computer power required.

One of the advantages of the eddy-resolving model is that it tends to be more realistic than models working from the mean, the time-averaged. It is now fairly clear that vortices of varying characteristics are at work throughout the ocean, dissipating energy through friction, mixing waters and rearranging living things and their food and such properties as heat and salinity. Eddies snap off the Stream, its extension, and other currents, or they can arise from smaller-scale instabilities. "Eddies are the way the ocean works," Holland believes. "They control the large-scale features of [global] circulation, including surface current systems and flow in the depths."

You can do two things with a numerical model. You can start with the best physics you have for a patch of water and translate it into algorithms suitable for supercomputers. Then you start things up and see how long the physics mirrors reality. The two won't stay synchronous long, Holland says, because a turbulent system drifts away to some other random state. But to the extent that they do, you know your science is suited to the phenomena it seeks to describe. The other thing you can do with your model is to optimize reality at the expense of your physics. You run a predictive model, one frequently corrected by feeding in information from ships, moorings, and satellites. Both approaches have their uses. Holland says he has an excellent model of the Gulf Stream and its meanderings, "but the meanders aren't going to be in the right place at a particular date."

No model is completely accurate. There are always gaps in the physics. Oceanographers still don't know much about the actual effects of evaporation and precipitation on ocean circulation. Then there are events like mixing and overturnings at scales well below the model's resolving power. Their effects are also a matter of approximation. They are parameterized, in the modeler's jargon, lumped together and assigned a best-guess effect. The wind is far from precisely measured at sea, and the errors creep into atmospheric models Holland uses to apply wind stress to his sea surface.

Still, the models continue to get better. Their translations and sophistications of Newtonian law are sometimes so good that they can produce oceanic features that decades of observations have missed. Since the means of observation are also improving, a remarkable synergism has developed between the observers with their proxies and the modelers with theirs, a three-legged race for reality. In the mid-eighties Holland needed hundreds of expensive computer hours to run some of his models. But computer evolution is in a frenzy. It is probable that machines many times more powerful than the giants of the mid-eighties will be available by the mid-nineties. It is also possible that by then new models and new ways of getting and interpreting data will be generating information in amounts that will satisfy if not satiate the new molochs.

Even now, models exist that track certain interactions between air and sea, despite the difference in requisite scales of resolution. The coupled system is beginning to show how the Gulf Stream meandering process redistributes heat in the ocean and is suggesting how that heat feeds back into the atmosphere. Models can now take satellite readings of sea-surface height and temperature and, by themselves, make a good stab at filling in the rest. Holland says that if the improvements keep coming, scientists will eventually have the ability to simulate the whole system.

Forecasting in the ocean is still a young thing. Holland does some of it, running his North Atlantic model and watching the larger features propagate. Another investigator, Allan Robinson, concentrates on somewhat smaller scales. He is convinced that he and his people can take the Stream, from Hatteras to the Maritimes, and tell a large and varied constituency what the axis, the meanders, and the rings will be doing weeks in the future. On paper, Robinson is pure Harvard. He went to college there, got his master's and doctorate there, and he teaches there. But a don he is not, though he has more than enough memberships and authorships to qualify. He is, root and branch, a scientist. And he is a showman: striking on his feet, a face that is almost classical, silvering hair. He is a driver, about as good at getting what he wants from Washington or other helpful places around the world as you'll find. He lives at a sprint, and I've never seen him winded.

Robinson has assembled a team of youngsters and put them to work in basement offices below his. He has arrangements that give him advanced interpretations of satellite data and advanced supercomputers to turn them into simulations of the Stream. For a couple of decades, Robinson has been working in regions of

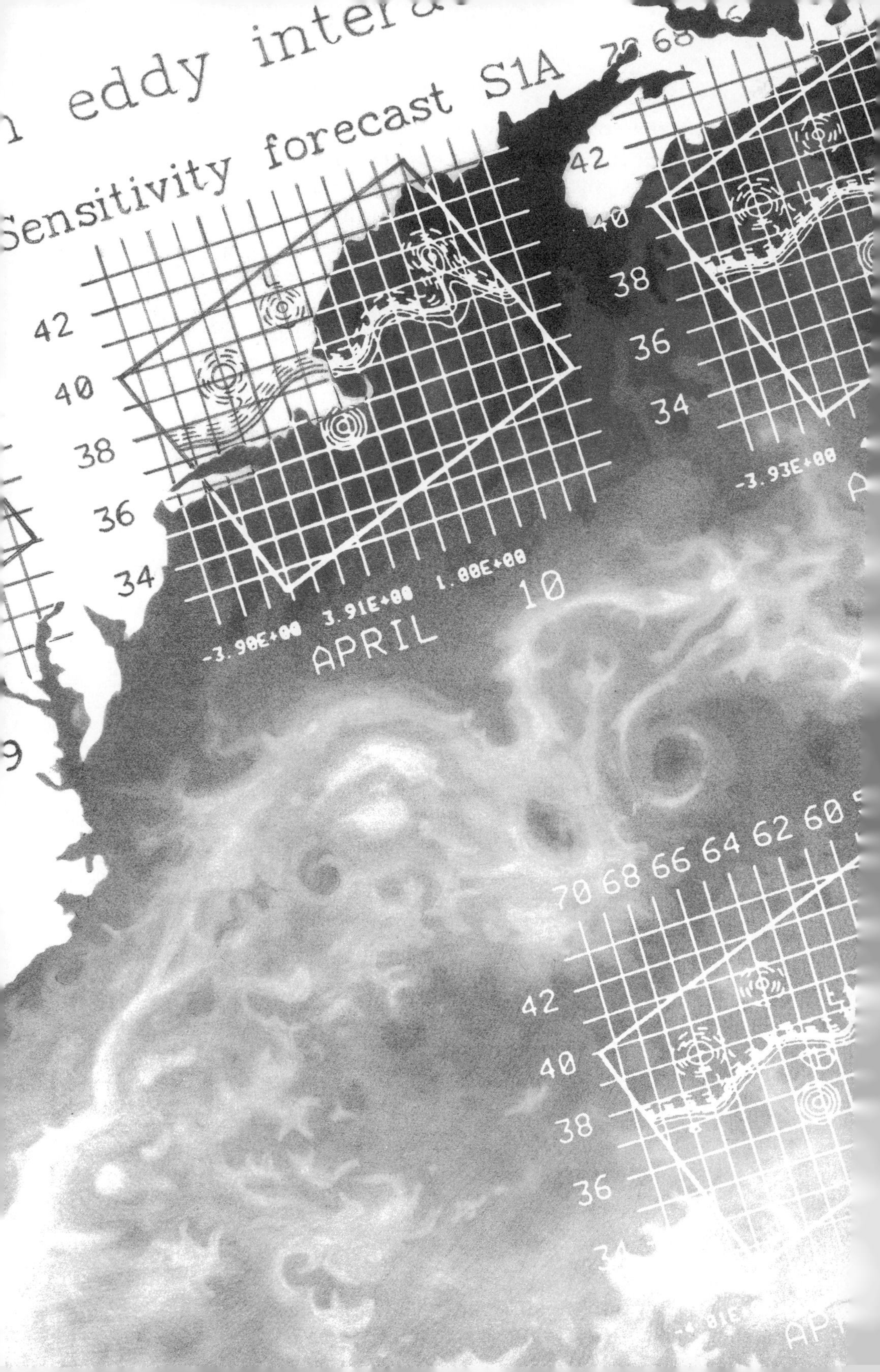
eddy
Sensitivity forecast S1A
42
40
38
36
34
-3.90E+00
3.91E+00
1.00E+00
APRIL
10
42
40
38
36
34
-3.93E+00
70 68 66 64 62 60
42
40
38
36

particular dynamical complexity. He was involved with huge international experiments studying eddy activity in the open ocean. That and related work led to the development of a coarse-grid model of the North Atlantic in which Robinson embedded areas of ocean where his group had sufficient data to permit use of a finer grid. The Gulf Stream was one of them. There were plenty of answers still to find, he told me three years ago, and a lot of very detailed models to run in the next few years, but, he said, "I really believe we are going to solve the major dynamical input of the Gulf Stream system to the general circulation problem." And, he added, "You're going to love it."

I checked back with Robinson and some of his team every few months. Each time there was something new to show, some further effort to educate me in this dynamical property or that technique of computer handling. I learned more about the errors Holland had mentioned, and came to see error as something to be managed rather than deplored. Error could be measured, just as scientists measure oceanic properties. The Robinson group uses a process that reduces error by looking at information from several angles. They try to keep the model running on what they know and use statistical estimates to get at what they don't. They constantly test the model for its sensitivity to this and that, for its robustness, for its relation to reality, and keep testing, validating.

Casting is the common word in the group's lingo. The modelers hindcast — that is, go back in time to an interesting period rich in accumulated data from hydrographic instruments and satellites and see how long the model follows the figures. They nowcast, bringing every bit of data they can acquire to bear on giving the model a full description of how things are at point go. Their scale of operations is tens to hundreds of kilometers across sea surfaces, days or weeks in time, the scale of weather in water as against the thousands of kilometers and hours and days of weather in air. "If you want to know what the weather will be," Robinson says, "you've got to know what it is." The modelers then forecast by allowing the model to run forward. That is what

I loved, watching the zeroes and ones of binary code take over the turbulence racing along beyond the continental slope and move it through the days to come — watching experts learn.

Physics differ. The physics of the equatorial sea is different from that of the northern Sargasso. The physics of the surface is different from that of mid-depth, from that of the bottom waters. To do its work, the Robinson group has developed a kind of automotive gear train of models, each version coupled to the rest in a sequence that saves thousands of hours of programming and running time. When they have no data on a Stream feature — an eddy, say — they plop in a dummy, a ready-made model of a stereotypical eddy that will do the job until more information arrives.

Flexibility like this has come to be essential to their artifice. They are working, after all, with a domain, a selected block of water astride the envelope of the Stream. The currents and rings passing through the domain are so dominant that they suppress the influence of the smaller whirls and flows that so perturb quieter oceanic regions. That is something of a boon for the Harvard modelers. Still, the modelers must know when a meander is coming down on them from the Inlet off Hatteras or when various thrashings downstream of the domain should be taken into account. To do that, they add boxes to their box, one each at the assumed points of inflow and outflow. When they get news of intruders, either from the larger basin model or from satellite images, they forecast what will happen in their entry and exit boxes.

Eventually, of course, the model, like Holland's or any other numerical simulation, takes leave of its senses. Nonlinear interactions make it possible for errors occurring in scales not of interest to the researcher to move into scales that are. Neither man nor machine yet knows enough to stay on track very long after that. The physics in the model can still produce phenomena of interest to theorists and the like, but for other purposes the model needs fresh food. Over most of the time I visited the group, it relied on a diet of infrared images from National Oceanic and Atmospheric Administration satellites — the same birds that supply Otis Brown and Peter Cornillon and Jenifer Clark. The Harvard team also got help from Brown and others in data interpretation. To connect sea-surface temperatures to what was going on below required different instruments. The AXBT, an airborne version of the expendable bathythermograph, seemed particularly useful. It carries a radio that remains on the surface as the instrument sinks, telemetering temperature-with-depth information to the aircraft that dropped it.

Robinson didn't have to spend too much time looking for AXBTs. The supply he needed came to him — in uniform. Naval warfare is increasingly undersea warfare; the submarine is the capital ship of the Soviet navy. The missile-carrying submarine is a major deterrent for both superpowers, and both spend hugely of time and money to keep track of each other's deep runners. Two hot wars and one long cold one this century have provided more than enough in the way of field trials to bring antisubmarine warfare to impressive levels.

Sound remains the most reliable medium for stalking, particularly in the deeps. There, close to the SOFAR channel more than a kilometer down, a good bang produced off New Zealand can be heard off Bermuda. Nuclear submarines make excellent platforms for listening devices. They are also noisy at speed, though many of their earlier swishings and clanks have been damped by new materials and design techniques (or masked by the ever increasing incidence of noise in the sea from shipping, oil rigs, and other human endeavors). As long as they make any appreciable sound, subs can be tracked acoustically, from ship-mounted hydrophones, from killer submarines, from sonobuoys dropped by aircraft, from listening devices planted on the bottom.

But not always. Well before World War II, the U.S. Navy knew that on certain afternoons in the tropics its hydrophones went deaf. Woods Hole's old "A-boat" sailed down to the Caribbean to investigate, and her people were able to demonstrate that the warm mixed layer near the surface created a sound shadow strong enough to hide a wolf pack. In time, it became clear that wherever there were rapid changes in temperature, vertical or horizontal, sonar operators would have trouble. Sound ducts or layers of anomalously warm water can form on the surface, and a skilled submariner can use them to avoid detection from destroyers above or enemy subs below. The ubiquitous eddy makes a lovely hidey-hole, and so do the edges of strong currents. To the military, the water in those places is bad water.

For both the Soviet and the American navies, oceanography has become enormously important. I went to a workshop in 1986 attended by an interesting mix of academics and officers where one of the latter quoted from an enemy oceanographer: "It is my responsibility to ensure that every Soviet submarine understands the ocean and how to hide in it." By implication, it is the responsibility of the U.S. Navy to understand the ocean, how to hide in it, and how the other side hides in it. We, like they, have our military oceanographers. We have the Office of Naval Research, which underwrites a large part of oceanographic work in this country. We have satellites operated by the navy.

Robinson learned that some of the navy's patrol aircraft were trying to make sense of the Stream, where bad waters abound, by dropping AXBTs, and they were having problems. He and the navy worked out an arrangement whereby some patrols would fly routes requested by the modelers and send their AXBT data to Cambridge, where Robinson and his group would use the data to correct the model and increase its accuracy. Copies of the forecast would be forwarded to the navy.

After one of the first cooperative exercises, I looked at representations of the Stream produced by satellite data and by the model. They looked like cartoons of weather maps. I saw outlines of warm-core rings inshore of the main current, cold-core rings offshore, the Stream axis, and a great "deep-sock" meander.

Robinson explained that one of the toughest tasks was to identify the domain that would be most useful to his clients. The navy was interested in a piece of ocean that was too small for effective forecasting by the model. The Harvard modelers chose a larger area, one in which meanders and the rings interacted dynamically, the vorticity or spin of one influencing that of another. The model wants a dynamically happy ocean, Robinson said, and the chosen site met that need. His group was able to show the navy that working with the larger domain would result in usable forecasts of Stream activity in the navy's patch of water: if you want to know what will be going on there tomorrow, he was able to suggest, drop your AXBTs here today.

Each time I went back, the Gulf Stream project had grown. The group acquired a scientist-manager for the part of its program it now called the Harvard University Gulfcast. He once worked for Shell Oil, trying to predict when warm-core rings would swirl through its deep drilling site — the same one Peter Cornillon studied. The manager used to fly out in a small aircraft and drop AXBTs, so he knew the instrument well.

Every week the group prepared another Gulfcast. They first took a look at the previous week's forecast, then matched it against the latest infrared satellite data. If there were questions, and there often were, about the precise location of this ring or that meander, they asked the navy to drop bathythermographic arrays on the question marks. They then inserted their standard features — eddies, currents — and pushed the button. Out came the forecast, day by day, as the model puttered and fretted and filled in the voids.

Some oceanographers I talked to are not convinced that Gulfcasting will turn out to be all that useful. "Hell," said one scientist, "I could run a meander down the Stream for a week with my pocket calculator." I asked Robinson about that. He had just been showing me a Gulfcast that predicted an unusual phenomenon: the birth of a pair of interactive rings. There was the starting point, showing a most un-Gulflike meander, and there was the result, verified by data. What about that claim for the pocket calculator?

"That's bullshit!" Robinson snapped. Then he calmed down and riffled through his printouts. "They might be able to run a meander with a calculator, but they can't make that double ring."

Under conditions that are ideal — an oxymoron where fluids are concerned — that part of ocean movement due to internal dynamics might be predictable, theoretically, as far as four months in the future. Of course, that would require a global observational network for the oceans that doesn't yet exist. Robinson doesn't feel he has reached any wall yet. On occasion, when his group has tested the model against past data in research studies, it has followed reality for as much as a month. He feels that simulation is one of the greatest advances in the scientific process in centuries, joining theory and observation as an equal partner. But he is cautious about how to proceed. Unlike some computer experts in the navy and elsewhere, he thinks it is premature to concentrate on global modeling. For him, regional modeling offers more for the present, particularly since it is both flexible and portable and, when the time comes, can nest well in the larger scales — eddy-resolving or ocean basin, and global. In every case, simulations will serve specific research or practical interests. "There is no such thing," Robinson once told a workshop, "as a general-purpose oceanic model."

The Harvard Gulfcast model was conceived as a research tool and serves as one. It is making runs during the SYNOP program, the multi-instrument analysis of the Stream, and Robinson has been urging instrumentalists to experiment with more real-time telemetering from their devices. He is now getting altimetric data from the navy's Geosat satellite, and that should help. So should the work his group is doing in developing so-called primitive equation models that can accept more physics than can other types of Gulf Stream models. The Harvard group is also expanding its research territory, to areas around Iceland and to the northeast Pacific.

The Harvard models clearly have lay potential. Anyone who

needs to know what the ocean looks like should want to know what the ocean will look like. Not much has been done along those lines as yet, but the group is in close contact with Jenny Clark. It even sent an emissary to join her in briefing some blue-water yachtsmen on where the Stream would be located during an upcoming race. True for the sailor, true for the scientist. "If you can't predict," Robinson says, "you can't understand."

9

Northeast Passage

THE WINDS AND WATERS of the North Atlantic subtropical gyre produce stunning weather, none more so than the hurricanes born in Africa as depressions and steered westward by the trades to bring heat and moisture and destruction to the Caribbean and beyond. Columbus weathered one. He was coming into the port of Santo Domingo on his last voyage. The place was jammed with some thirty ships and caravels about to leave for Spain, and the Admiral was denied entrance. Bartolomé de Las Casas, chronicling those times, says Columbus sent a message to the commander of the fleet saying a terrible storm was on the way and advising him to hole up. He himself pushed on a few leagues west to safe harbor.

The fleet sailed. "And after thirty or forty hours," Las Casas related, "there came such a rare tempest and so violent, that in many years men sailing on the seas of Spain and in other areas had not seen one equal, nor so sorrowful had they experienced. Twenty ships perished with the storm, without any man, small or great, escaping, and that neither dead nor alive could be found." Santo Domingo was flattened. There, "it seemed as if all the army of demons had escaped from Hades." Columbus fared well, refitted, and sailed on. José Carlos Millas, who plotted the courses and intensities of early Caribbean hurricanes, rated this blow as a heavy gale. Its winds, he thought, didn't attain the minimum velocities for a hurricane of at least seventy-three miles per hour, or force twelve on the Beaufort scale.

Heroes of the Conquest witnessed and survived true hurricanes. Ponce de León commented on one that carried people through the air. Cabeza de Vaca, before he went wandering in the American southwest, wandered through a hellish night in Cuba, hearing "a great uproar and the tinkling of little bells and of flutes and tambourines and other instruments," and the following day came upon a ship's small boat on top of some trees and "two men from my ship . . . so disfigured by striking against rocks as to be unrecognizable." The true killer occurred in 1780. It came in slowly from its nursery somewhere far to the east in the tropical Atlantic, and it killed an estimated 22,000 people on Barbados, St. Vincent, St. Lucia, and Martinique. It drowned 4,000 French soldiers in a Caribbean convoy and veered around toward Bermuda and sank British warships sailing for home.

Hurricanes can have an effect on the Gulf Stream. Millas quotes an account describing a "heavy gale of wind from the northeast" that "so greatly impeded the current of the Gulf Stream that the water, forced at the same time into the Gulf of Mexico by the trade winds, rose to such a height, that . . . the Tortugas and other islands disappeared. . . . " Millas figured the cause was a storm center stalled over central Cuba. Today meteorologists using satellites can see strong hurricanes tearing up the Sargasso, spinning cold water up from several hundred meters down and leaving it on the surface for weeks as a wake. The surface temperature patterns of eddies shed by the Loop Current in the Gulf of Mexico have been all but destroyed by the same process.

The effect of the Stream on hurricanes is another matter. There is not much difference between sea-surface temperature and Stream temperature during the hurricane season, not enough for storms passing over to sense. Perhaps one veering north along the Stream's track later in the fall could benefit from the heat anomaly. Generally, hurricanes approaching the Atlantic coast tend to weaken faster than those approaching the Gulf coast, probably because they get a sudden flood of cold water under them as they pass over the north wall of the Stream and into continental slope water. A lot depends on the speed of their movement over the water and the location of the feeder bands that draw in air from up to a hundred miles out.

The Stream's strongest links to weather are more localized and arise from the influence of its warm core on the air above — chimney storms, for example. Ample supplies of very warm water under a flock of thunderheads can help generate an updraft that surges high into the clouds, displacing cold air. In the ensuing turbulence the dense, cold air drops down, gaining speed as it goes, and smashes into the sea surface. Winds spread out in all directions, some at gale speed or more. The whole thing triggers in minutes and lasts about as long as a squall. It is too small to be forecast by standard methods.

Chimney storms or microbursts aren't limited to the Stream, of course. It is probable that one took the famous topsail schooner *Pride of Baltimore* north and west of Puerto Rico in May of 1986. A ship rigged like the clipper *Cutty Sark* recently fell victim to a microburst near the Stream, off North Carolina, and another square-rigger went down off Bermuda not long ago. High masts and abundant rigging seem particularly vulnerable to microbursts, possibly because the winds can come in from a high angle off the sea. Then there are the craft never heard from again. Some young acquaintances of mine sailed out of Woods Hole a few years ago bound for warmer seas. No one has seen them since. Who knows? Stories like this are common enough. Those who relish mystery invoke the killer myths associated with the Bermuda Triangle. Scientists think meteorology plays more of a hand than mystery.

A storm that causes far more damage than chimney storms is what the Nova Scotians call the winter hurricane. Meteorologists call it the bomb, or a rapidly intensifying cyclone. Satellites can spot its signature, a comma-shaped cloud mass. Bombs have a habit of passing northeast near the heavily populated corridor between Washington and Boston. That stretch of sea has some of the worst winter weather in the world, but not every extratropical low dropping sleet and snow there turns from a nuisance into a nightmare. Those that do — and usually there are several of them each year, most in the months of January and February — suddenly go crazy. Atmospheric pressure at their center drops radically and rapidly, often in six hours or so. The bomb can develop an eye like a hurricane and winds of hurricane strength.

Most of us have heard about the really bad explosions, like the ones that immobilized Boston in the winter of 1978 or damaged the *Queen Elizabeth II* that same year. The storm I remember was a rare double bomb in February 1982. It went through two barometric plummetings in twenty-four hours, and in the middle of the second one capsized the huge oil rig *Ocean Ranger*, drilling for Mobil in the Hibernia field off Newfoundland. It killed all eighty-four of her crew. The reason I remember this is that the next day I went out to another rig chartered to Mobil, working over Georges Bank. I saw roughnecks in mourning.

Bombs are found near western boundary currents — the Gulf Stream, the Kuroshio in the western Pacific, elsewhere. They do their work in and near shipping lanes, areas of strategic interest, fishing grounds. The U.S. Navy received enough reports of damaged warships that it, along with the National Science Foundation, is helping fund much of the current research on explosive storms. Scientists using patrol planes, ships, buoys, and satellites have found that the storms often have their beginnings in the southeastern United States, incubate off the Carolinas, and go wild off New York, New England, and the Maritimes.

What appears to happen begins with the arrival of extremely cold Canadian air over land east of the Appalachians and adjacent coastal seas, brought down by winds circling around a high over the Great Lakes. Air given extra warmth and humidity in its passage over the Stream presses in toward the land. The arctic air cannot retreat westward because of the mountains, so the warm flow vaults over it, and instability ensues. Precipitation from the cooling warm air falls through the Canadian mass, and things begin to move in a vortex, counterclockwise. Although the bomb is still too small for weathermen to spot easily, it is now armed.

It will be years before meteorologists know enough about tricks of explosive intensification, including the role played by vertical air movements, to develop the physics necessary for predictive models. And years before oceanographers understand what storms with these characteristics do to the Stream system. There is evidence that winds blowing along the Stream axis affect its lateral motion. What happens when a miniaturized hurricane passes through? And what is the relationship between the disposition of the Stream, its meanders and eddies, and the personality of a developing bomb? Researchers trying to answer that one are in contact with Allan Robinson and his Harvard model.

The farther down the Gulf Stream you go, the worse the weather gets. The stretch of the North Atlantic off Newfoundland causes oceanographers to wince when they think about it. The Stream comes coursing around on a bow bend, close by the Grand Banks. The Labrador Current comes down inshore, from cold and narrow Davis Strait, and closes with the Stream. All is confusion in the resulting mix: fog and cold — and great fishing. Some of the heat brought north by the Stream ends up in an eastward flow composed of the relatively narrow North Atlantic Current, running as something of a jet along the front formed by the edge of the subpolar water, and a more diffuse movement of the sea under the westerlies, the North Atlantic Drift. Temperature differences in this mixing bowl, particularly where the Stream and the Labrador Current jostle each other, are greater than just about anywhere else in the world ocean.

Ice is a complicating factor here, as Art Snyder learned in 1985 when he sailed *Welcome* on the northern route heading for western Ireland. From there he would sail toward the Hebrides and then ease on down the British coast, cross over to France, to Spain and then leave *Welcome* at Vilamoura. They crossed the Grand Banks in June. The Canadians had told them they'd see ice right on their track, and they did.

Art's daughter Carrie had the watch. She couldn't see much in the foul and foggy weather but thought she had a big ship on the radar. Snyder kept checking, noticed it wasn't moving, and figured it was a monstrous berg. "I didn't enlighten anybody about it," he told me when I saw him in Boston some time after my own voyage on *Welcome*. "Icebergs aren't dangerous to a boat like mine, traveling four knots, because you can hear them and smell them, and you're not going to run your bowsprit up on them." Snyder was leery of the smaller chunks, the bergy bits and growlers. Fogbound, they came within fifty yards or so of something "as big as a whale out of the water, so ten times as big as a whale under the water. And that would have made one helluva mess out of us if we'd run up on it." They got through all right, but not before they just barely missed collision with an American ship towing sonic gear.

Icebergs float down with the Labrador Current and its outriders. Some drift right into the Stream, which melts them in short order. Some ride in the shipping lanes. One of those, back in 1912, sank the *Titanic*. That sinking resulted in the establishment of the International Ice Patrol, a service operated by the United States Coast Guard for the benefit of all shipping transiting the region. In the spring of 1987, I flew with the patrol.

The crossroads of the world: that's what Gander, Newfoundland, called itself when I landed there a few years after World War II. Propeller planes flying the North Atlantic from places like Shannon airport in western Ireland refueled there before moving on to mainland North America. Now the long-range jets fly direct, and the crossroads stands empty, a great expanse of concrete next to a small town on a large island. The eastern Europeans use the airport, and the Cubans, and sometimes the Soviets. One of their planes came in during bad weather a few years ago and made what a witness described as a politically proper landing — to the left of the runway. Another came to help it and also missed. After a month or two, the Russians sorted things out and went home.

The International Ice Patrol flies out of Gander as of this writing. The C-130 Hercules patrol craft come up from North Carolina for deployment. They sit next to the Canadian ice patrol aircraft, which look more for sea ice that can jam harbors than for the deeper-draft bergs. The Americans fly seven- or eight-hour flights, making carefully preplanned transects of a stretch of ocean off Newfoundland covering ten thousand square miles. They look for bergs, visually and with proxies, and they report sightings, along with wind and current data, to ice patrol headquarters in Groton, Connecticut. Groton runs two computer models. One predicts berg travel and therefore limits of berg penetration into seas traversed by maritime commerce, including vessels of the patrol's twenty participating nations. The other estimates rates of deterioration due to melting and, more important, wave erosion. Waves of a certain height create resonance within a berg that eventually fractures it.

The Hercules is a flying warehouse, the next safest thing, I was told, to staying on the ground. That may be so, but it also has to be the next noisiest thing to a tank. Earplugs cut some of the racket from the four turboprop engines and the protestations of the hull, but enough survives to pester crewmen unmercifully. Our aircraft was equipped with rows of seats set as if we were going to watch a movie. People with nothing to do at the moment sat in those seats and listened to Walkmans. One needled away at a crewelwork tiger. Meantime, the sonics hammered away on their psyches.

"After seven, eight hours, we all turn into pumpkins." This from a lieutenant (jg.) named Neal Thayer, senior ice observer, a man with a mustache, a degree in biological oceanography, and three years of learning a lot about what currents do to and with ice. Bergs are calved in the thousands from glaciers, most of them on the western coast of Greenland. Those born on the east side generally enter the cold East Greenland Current, which merges with the warmish Irminger Current to form the West Greenland Current, which flows along the coast into the Labrador Sea, losing out to the counterclockwise current there. After a time the West Greenland Current strikes across to the west side of the sea to run south, parallel with the Baffin Island Current, the twin ribbons forming the Labrador Current — which, as we know, runs to play with the Stream in unfathomable patterns.

Icebergs, Neal Thayer said, are metamorphic rocks that happen to be made of water. Their ice was formed under pressure, so they last like rock and fracture like rock. In a highball glass, berg fragments fizz a little as they melt. Thayer brings some bergy bits home from time to time to amuse the uninitiated. I read in the *Washington Post* that in Tokyo, where people are beginning to dip into their savings for such sillies as cans of mint-scented oxygen, the price of ten cubes of arctic ice guaranteed to be ten thousand years old was $7.50 as of the end of 1987.

By definition, icebergs are at least seventeen feet proud of the water and fifty feet long. Smaller than that, they are growlers. Berg sizings run from small through medium and large to very large — at least 240 feet high and 670 feet long. That is how ice observers report them, on the rare occasions when they can get a visual sighting. The biggest Greenland berg reported by the Coast Guard hulked 550 feet above the sea.

When I first talked with Stephen Osmer, a lieutenant commander who runs the patrol, 1987 had looked like a heavy ice year. But by March, severe storms had changed things around, driving the sea ice right up against the coast. They blew the bergs west, Neal Thayer said, "or turned them into margaritas." The hunting might not be so good.

We flew patrols to check oceanographic conditions and patrols to check ice limits. The pilot baked banana bread while people talked over the intercom, long rambling discussions about the intricacies of the airplane, about hunting in Alaska. The theory was advanced that a light rifle was just the ticket for hunting Kodiak bears. You sight one, you shoot your buddy in the foot and run. On and on, while the plane followed its courses, turning just so, quartering the sea to give the Side Looking Airborne Radar the best views.

The SLAR on our patrols sat like an altar well forward in the huge, light green bay near the observer's windows. At our altitude of eight thousand feet, its beams reached out about thirty miles on each side of the Hercules, striking the sea, scattering, returning with information to be translated on a screen and a printer. The priest of the altar, a canny noncom, read from the mass of packed lines and announced this spike to be a berg, that a ship. Like a satellite scatterometer, SLAR can read the roughness of the sea skin. Roughness is associated with wind, and in this region of extreme temperature contrasts, cold dry wind from the continent blowing over warm water can produce turbulence in the boundary layer that is radar-bright but that shows dark in the SLAR negative readout. Ergo, dark patch equals warmer water — at times.

Since the SLAR also penetrates clouds and the eternal fogs of the banks, Neal Thayer and his associates have built a fair record of the positions, spotted this day or that, of shifting oceanic fronts below them. Thayer showed me one front, tracing a crack angling through the mass of parallel lines. "That's probably an edge of the Labrador Current," he said. We both wore headsets, the only way to hear anything but rumbles. Trying simultaneously to work a sandwich between the struts and wires in front of my face, I nodded and got mayo on the mike.

The year before, Thayer had collaborated with two colleagues on a paper discussing a front encountered well offshore of the Banks. Thayer et al. pointed out that SLAR was not as good as infrared at defining a front, but that in the entire three weeks of the patrol under consideration, there was exactly one day clear enough for the thermal sensors of a satellite to function. One of the paper's authors worked aboard a participating cutter; the others flew. The shipboard data agreed with the SLAR images and with the tracks of some surface drifters thrown in for good measure. The investigators were able to show that during their patrol period they had been looking at a warm-core eddy shed by the North Atlantic Current. More important, the eddy and its clockwise motion had apparently teased a portion of the Labrador Current into leaving its customary flow along the eastern edge of the Grand Banks and heading out to sea to parallel the North Atlantic Current.

On my patrol, the bergs were scarce indeed. We saw none on the first couple of flights, though the SLAR showed an occasional spike. One day, when the plane was due to head south into less interesting water, I drove along the eastern bays and down long peninsulas to poor and lovely Newfoundland fishing villages. Grounded bergs were there, a couple of miles or more offshore, in their remove as natural as the headlands reaching for them. The next day we flew low over a thin, ribbed fog, looking for bergs. I crouched by the starboard observer's window, watching the shadow of the plane skim the mist in the center of a sun halo. A great alien appeared below me, a whiteness that was frightening, a sudden resolution of shape in the shapeless fog. It slid aft in the instant and left the visceral turbulence I felt when I stalked my first deer. That wild.

The fog thickened, and the pilot took us back to the coast and over some grounded bergs. He opened the great ramp in the tail, and we saw diesel exhausts attenuating behind us and the ocean two hundred feet below. "Fifteen seconds," the pilot radioed to us. "Five." I thought I felt a bump, the cold over the berg, and then it showed. The flight deck had been giving me a little razzing about journalists always describing bergs as "a beautiful shade of blue." I looked especially carefully and saw white I had never seen before. A fortress, this one, bastioned. Where the tip joined the greater mass below, waves had cut an emerald beach. Another was a tabletop; another, trailing growlers and bergy bits, was on its way to becoming a margarita. The pilot pulled up sharply, and we angled off toward the crossroads of the world. Neal Thayer told me that every year on the anniversary of the *Titanic*'s sinking, a Coast Guard reconnaissance plane flies over her grave and drops a wreath. The spot lies some 340 miles southwest of Cape Race, the seaward tip of Newfoundland. Sometimes the wind catches the wreath and throws it back like a Frisbee into the plane's bay.

A couple of months after my time with the ice patrol, I arranged to ride the top of the North Atlantic subtropical gyre as I had ridden the bottom. The ship of opportunity was somewhat different from Arthur Snyder's command. Eight hundred and sixty feet long, a roll-on, roll-off container ship, she could carry a couple of thousand *Welcome* hulls. She was the *Barber Tampa*, Ole Abrahamsen master. Built in South Korea in 1983. Nationality, Norwegian. Port of registry, Tönsberg. A few years ago, while writing an article on the pilots of New York harbor, I had ridden *Tampa*'s little sister in to her berth in the kills behind Staten Island. We went in at dusk, and I remember the vault of the bridge, dull red in the night lights, and, beyond the wall of windows, the jeweled city.

I caught the *Tampa* at St. John, New Brunswick, a pretty town on the Bay of Fundy. She had come up from the Panama Canal to Mobile, Alabama, and from there right up the east coast to St.

John. She and her siblings serve three huge markets — North America, Europe, the Far East and Australia. Different crews take them around the world, carrying yachts, cars, earthmoving equipment, once some American M-60 tanks bound for Kuwait. At St. John, paper and lumber were being loaded over the great ramp on her starboard quarter by maniacs driving John Deere tractors hauling long flatbed trailers.

We left the first of August, 1987, rounding the tip of Nova Scotia and running a little south of the great circle route to the Isles of Scilly and then up the Channel and the Thames to the docks of Tilbury, near London. The trip would take seven days and some hours, and we would steam at between nineteen and twenty knots, more than three times *Welcome*'s best speed.

Captain Abrahamsen had checked ice conditions. There were a couple of bergs around the Tail of the Banks, the seaward limit of the Grand Banks fisheries that run four hundred miles or so east and a bit south, toward the Azores. *Tampa* encountered no ice, not much even in the way of weather. We forged along, as much above the sea as on it. Abrahamsen said the ship didn't take kindly to foul weather coming at her bows. She had too long a spine to go full out, and in those conditions he eased her off accordingly. Eighteen men to run her, six to a watch: six souls to manage a giant. I have never felt so alone as aboard *Tàmpa*. There was always someone up on the bridge, but elsewhere it was ghostly. Solitude even pervaded the mess, where we ate excellent food in the murmuration of Norwegian vowels, slow and deep.

Abrahamsen and I talked on the bridge sometimes after dinner. Sure, he worked with the Gulf Current all the time. He caught it coming up from the Panama Canal, and he rode it through the Florida Straits. It would give us a bit of a lift even as we moved out into the open Atlantic. And, he said, it gave his country more than a bit of leaven in the hard winters. Abrahamsen comes from a group of islands on the east side of the fjord leading up to Oslo. He was apprenticed to the sea as a fisherman, drift-netting for mackerel and following the herring and shaping a talent for knowing the weather. Later, a boatyard asked him when they might expect good weather for towing a big hull to its owner. He ducked out of his house and looked around and after several such expeditions told them to go ahead. You might get fog, though, he said. They did, but had a good voyage. Did he speak to the Lord? they asked. No, he said, I just look around.

When I met him, Abrahamsen had been twenty-seven years a captain. He came to the big ships after he had pulled a tanker off the shoals in the Caribbean and received a certificate of merit for his efforts. "You do it right," he said, "and you have lots of friends. If something goes wrong, no one knows you." He did not seem bothered too much by his job. His daughter, young and pretty, was making the trip with him, and she had her fiancé with her — what a pleasure to hear the word again. Like everyone else, she spoke English, while I, the American, contemplated my lingual incompetence.

It was a marvelous place and time to read and to wander out in spirit over the old northern sea. I thought of those who may have been the first to come east over these swells, following *Tampa*'s track. Archaeologists have found the bones and tools of people who lived along the coasts of the north. They were great seamen and fishers, harpooning swordfish with points beautifully fashioned from chert. The datings of their fires and bones and the similarities of their artifacts have led to a hypothesis that these Red Paint people, these maritime archaics, may have followed the Stream and its extensions to northern Europe several thousand years ago. If they could take a swordfish, I thought, they could, just possibly, have had the knowledge and the luck to cross over.

I had Saint Brendan with me, a translation of the *Navigatio Sancti Brendani Abbatis* written in Ireland, around the year 800. And a book by a contemporary Irish geographer and adventurer, Tim Severin, who led a small group of men in the building and sailing of a leather boat to demonstrate that Brendan could have reached North America. They themselves did, barely. If

Brendan did, he beat the Norse by four hundred years and Columbus by twice that.

It is a wonder tale, the *Navigatio*, full of symmetries and redundancies, yet it sometimes settles on what may have been the actual. Scholars believe Brendan did live, that he was born sometime around the beginning of the sixth century. The calling among many of the hermetic monks of Ireland was to travel the northern seas for Christ's sake and to carry with them the scholarship that most of Europe had lost. They founded monasteries from Scotland to France. They were revered, despised, slaughtered, yet they managed to survive in their stone cells all along the coasts and the islands of the eastern North Atlantic. They probably reached Iceland in some numbers by 800 and Greenland a century later.

The story of Brendan's voyaging is of a search for the Promised Land of the Saints, peopled by souls and wondrous creatures. Brendan built a curragh, "a light boat ribbed with wood and with a wooden frame, as is usual in those parts. They covered it with oxhides tanned with the bark of oak and smeared all the joints of the hides with fat." And they sailed off "into the summer solstice" on a journey that lasted seven years and took them back to the magical places. They encountered an island where birds chanted and their wing strokes sounded like bells; a whale named Jasconius who swam off when the monks mistook him for an island and built a fire on his back; an island where hairy blacksmiths hurled fiery slag at them (perhaps Homer summered in Ireland).

And then, the prize. "As the evening drew on, a great fog enveloped them, so that one of them could hardly see another." In an hour, "a mighty light shone all around them." They found themselves on the shores of a fine land so vast they could not find the end of it though they tried for forty days. A youth appeared and embraced them and told Brendan he could not have found the Promised Land sooner "because God wanted to show you his varied secrets in the great ocean." The youth told

Brendan he would soon sleep with his fathers, and so it was. The monks returned home and, soon after, the saint died.

The saga tells of a region of whales, and that could be the waters off Greenland. It tells of the coagulated sea, and that could be the weedy Sargasso. The Promised Land is shrouded in fog, and where would that be? Newfoundland is not fruitful and set about with precious stones, as was Brendan's last discovery, but it fits in fogginess.

I kept looking at *Tampa*'s radio operator, a young and obviously strong man with red hair and beard and a face at once fierce and sentient. Not much tampering with the Norse blood down the generations to him. He could have been in the broad *knarrs* with Eric the Red and his son Leif, sailing west from Iceland to Greenland and on. Samuel Eliot Morison wrote that the Norse, like other mariners, used latitude sailing in their explorations, departing their coast at the presumed latitude of their target and then maintaining that latitude as best they could, with a crude "sun shadow-board," with an eye on the direction of the swells, with luck. They could island-hop, from the Faroes to Iceland to Greenland, and escape long voyages on the open sea. And at times they may have been helped by what is called the arctic mirage; stable air sitting on ice behaves like a lens, bending light, causing objects over the horizon to loom above it.

They found Vinland, fruitful as Brendan's last find. Americans looked for it in New England and thought they had found a Norse tower in Rhode Island. But the best evidence for the existence of such a colony is L'Anse aux Meadows in northern Newfoundland, where a Norwegian archaeologist dug and came upon the sites of great houses such as the Norse used in Greenland, with a steam bath and a large cooking pit, and charred roof timbers dating back close to the year 1000, when Leif Ericson and his band, as the tales tell, came ashore.

No grape could possibly have grown at L'Anse aux Meadows, even though the climate was milder back then. But Morison cites authorities to the effect that "grape" was a bad translation: "*Vinber*, the word in the [Norse] sagas usually translated "grapes," really meant "wineberry, which might be the wild red currant" or some other lesser fruit. Morison's hunch is that as his father, Eric the Red, had dressed up his discovery of a drear coast by calling it Greenland, so Leif had invented vines in Newfoundland. Promotion is not a modern creation.

The Norse did much of their sailing on the subpolar gyre, the cramped and crooked circuit in the water north of *Tampa*'s track. This gyre turns to the left. The water cannot turn clockwise, as its southern neighbor does, because the winds drive it around counterclockwise. All along its course it is confronted with the skirts of continents and islands, with undersea rises and, on the south, by the moving water of the stronger Gulf Stream and the North Atlantic Current.

It is the devil's job to get data from the subpolar gyre in sufficient quantity and of sufficient quality to satisfy oceanographers. "O Greenland is an awful place," the whalers sang,

> A land that's never green.
> Where there's ice and snow,
> And the stormy winds do blow,
> And daylight's seldom seen . . .

Hugh Livingston, a Scot working at Woods Hole, follows currents with chemical tracers, some of which, like tritium and cesium, have been around since the first atmospheric tests of nuclear bombs. A fresher injection of cesium in the northeast North Atlantic has come from discharges of low-level nuclear waste from a British reprocessing plant called Sellafield on the Irish Sea.

Livingston has followed the flow of Sellafield effluvium well up into the seas that lie in the basins above Iceland, the Norwegian on the east and the Greenland on the west. "I'm getting the Atlantic water labeled as it goes in," he told me. "I can trace it under the ice." Using the full suite of tracers, he and other chemical oceanographers can keep on tracing, up toward the pole, or around in the counterclockwise gyre north of Iceland, or down and around Greenland and off toward the Labrador Sea. The rivers of Russia, Livingston said, dump great amounts of fresh water into the Arctic Ocean. Much of that water comes across on the clockwise transpolar current, ending up as part of the subpolar gyre: One great river, cold and fresh. It warms up as it goes south, but it carries the same signatures. It is extremely high in bomb tritium, because its source rivers, being relatively shallow, do not dilute it.

In our sloppy societies, there appears to be no end to tracers. When I talked with Livingston in the summer of 1986, he was preparing to sample radioactive chemicals recently injected in the sea as a result of the accident at the Soviet nuclear power plant at Chernobyl. In Atlantic water, he said, Chernobyl would show up as a sharp spike in the minute signals of cesium and other radioactive elements. The spike would broaden out slowly over a year or two. Since Chernobyl cesium has an isotope with a peculiar two-year half-life, Livingston and his people anticipate little trouble distinguishing it from Sellafield effluent or bomb detritus.

Another Scot, Robert Dickson, based at a fisheries laboratory in England, believes he and his colleagues have been able to reconstruct the voyage of what he calls "one of the greatest dislocations of the ocean climate, certainly in our sector, maybe globally, in the present century." Others call it simply the Great Slug, a blob of anomalously fresh water that chugged right around the subpolar gyre and even down into some of the eastern filaments of the Gulf Stream system, taking fourteen years, from 1968 to 1982, to make the circuit. Dickson hypothesizes that the whole thing started with a polar high that settled down over Greenland, directing streams of northerly winds down the Greenland Sea. The winds drove arctic water freshened by ice melt south to form a pool on the surface. It was too fresh to mix with the layers below, so it sat there, froze in the winters, melted in the summers, and grew. Eventually there was enough of it to flow down the Denmark Strait between Greenland and Iceland and out along the subpolar gyre into the North Atlantic. Fourteen cycles of seasons were unable to smear it out.

Dickson's data are very gappy in spots. He says that in researching the Slug, he found that the measurements of the eastern North Atlantic made in the earlier part of the century were better than more recent ones. There were more ships out then, including Nansen's *Fram*. But when the Slug was en route, it seemed that every time it bore down on good sources of instrumentation, namely the ocean weather ships deployed across the Atlantic, the sources would be withdrawn for lack of funding. John Gould of the Institute of Oceanographic Sciences, Britain's leading marine-science center, thinks that if another anomaly started its rounds today, there wouldn't be enough equipment to track it properly.

Gould and Dickson are both interested in the northern seas not only because they are located in home grounds but because they furnish much of the bottom water of the world ocean. For most of its life, oceanography has concerned itself with horizontal movement of the ocean, basically the wind-driven circulation. But increasingly, scientists at sea are studying thermohaline movement, the vertical convections by which the oceans essentially stay alive. Water cools, grows dense, sinks, flows somewhere else, and then filters up toward the surface bearing fresh nutrients. The strange thing about all this is that there are only two known places on earth — two rather small places — that supply most of the cold water that contributes so much to the balances of the planet. One is in the Weddell Sea, in Antarctica; the other is in parts of the Norwegian and Greenland seas. The Greenland Sea is a stinting source. Its southern borders are sills across straits, and only the top fifteen hundred meters or so can flow over the lip and into the Atlantic. But the coldest of that water is sufficient unto its purposes.

I read about vertical circulation as *Barber Tampa* slid along in quiet seas more than fifteen hundred miles south of Denmark Strait. I had brought a copy of Henry Stommel's *A View of the Sea*, and as I checked my marginal notations I remembered how Stommel had talked to me about his theories. We sat in a hot summer afternoon at the edge of his field, and he used dinner plates from his kitchen to educate me. There wasn't just one gyre in the North Atlantic, he said, but several, one under another in an uncertain stack. Each gyre, each dinner plate, tilted so that its northern edge outcropped on the surface of the sea while its southern was submerged. Each plate lay somewhat to the north of the one over it.

Later, Stommel drew the path of a parcel of water on a black globe and talked it along the upper layers of the Sargasso, curving southwest and then up north and east in the Stream: "The water is colder now than it was at the beginning. And so as it goes around, it subducts under the upper layer, goes around, goes back up the Gulf Stream and comes out farther north." On each circuit the parcel makes a deeper traverse, outcrops farther toward the pole, until some of it flows into the Norwegian Sea. There, it cools to within only a degree or so above the freezing point of seawater (about −1.8 degrees centigrade) and then comes back along the western boundary of the ocean — the flow, hugging the North American continental slope, that might just as well be called the Stommel Undercurrent. Some of that deep, frigid stream detrains into the interior of the Atlantic, but most of it goes on to other water. "It's interesting," Stommel said, "that half or so of the bottom water in the Pacific and Indian oceans comes from this so-called Norwegian Sea North Atlantic Deep Water."

The *Tampa* was closing on Ireland now. In less than a day we would pass to its south, raise the Isles of Scilly, swing around into the English Channel, and head up the Thames. The great circle routes to the Channel entrance were closing on each other, and there was traffic all around us — tankers, container ships, bulk carriers. And, way out to starboard, a spike of bright sail, the ketch *Corazon,* inconsequential as a spent mayfly on a backwater of the Deerfield.

Corazon got on the radio and did what we had done aboard *Welcome* — got a fix from a passing giant. "Vessel to my port, do you have a weather report?" He was out there bucking a good northeast wind that we didn't even feel. *Tampa*'s mate gave what he could: a low lay off Portugal, so *Corazon* could expect a couple of days more of foul winds. "How are things in the Irish Sea?" The mate didn't know. "Please call me if you find out." The mate said he would.

In a half hour, maybe less, *Corazon* disappeared astern. I stood with the mate looking out over the high stacks of containers, over the bows, toward England. The great engine pulsed. I couldn't get *Welcome* out of my mind. Art, I thought. I'm crossing the sea in a skyscraper.

10
In Galway

THEY HAVE BEEN GARDENING here for a thousand years and more. There is word of roses paid as taxes in the thirteenth century. In the nineteenth, rich Irishmen imported plants from every continent. They built breaks against the west wind — the sea wind — and anything that could take the wet grew in glory. Plants escaped the walled gardens and struck out along the roads and lanes, the boreens.

I ended up in the west of Ireland, near Galway, looking from a car window at fuchsia bushes ten feet high flashing rose and garnet on the verges; at cabbage palm from Australia — a cousin of the lily gone giant; at a large-leafed invader from Chile claiming ground the way kudzu does in the American South. This is not the place to see the true delicacies of the subtropics growing far from their realm. The Isles of Scilly are that paradise, drier by far than the bogs and bald limestone ridges around Galway Bay. Still, roses bloom here on a good Christmas. There, the tenderest succulents thrive.

I had come directly from Tilbury, from *Tampa*, after the most expensive taxi I have ever ridden dropped me at Heathrow airport. From there it was an easy loft to Dublin and its National Botanic Gardens, where Charles Nelson, a botanical geographer, works. Months before he had sent me papers that introduced me to the sea bean and the legends of its wanderings. He took me out to the hothouses of his workplace and showed me vines propagated from a sea heart that had washed up on a western beach and then took me to that beach, called Dog's Bay, and beyond.

Like most Americans, I am too mixed in the blood to call myself anything but American or New Englander. But I cotton to my Scots line, and I know that my ancestors went to Scotland from Ireland. So, said I, looking out from Nelson's car at the Twelve Bens, three-thousand-foot hills that are the sierras of the Irish. So. I'm home.

We skirted Galway to go look at some botanical anomalies in the acid land of the peat bogs to the north and west. I could see part of the city at the head of the great bay. Nora Barnacle lived there, before she went to Dublin and married James Joyce. Columbus came there, a sailor in his mid-twenties. He wrote that he had sailed well into the northern seas, to Thule or Iceland. Galway was part of that Spanish trading network. "We have seen many strange things," the Admiral wrote, "especially in Galway of Ireland, a man and a woman of extraordinary appearance in two boats adrift." Finns or flat-faced Laps, according to Samuel Eliot Morison, or perhaps Greenland Eskimos. Columbus may have heard of Greenland when in Iceland. He might just possibly have picked up stories about Vinland to the west, but if so Morison could find no mention of it in his journals and postils, only the statement "Men of Cathay, which is toward the Orient, have come hither." And a note stating that contrary to popular belief the northern seas were neither frozen nor unnavigable.

Nelson takes a quiet pleasure in what he has come to know, and he shared it. I learned that *shamrock* simply means "young clover," and that's what it is. I learned that Ireland, like most islands, has very few species of plants that are unique to its lands. Most of what it has came to it from England and Europe after the last glaciation but before the melting ice filled the ocean basins and submerged the land bridges. Some species arrived later as seeds drifting in the wind or riding in the guts of birds, or, much later, as cuttings or plants or seeds imported by well-heeled horticulturists. A few plants — a very few — came by sea, by current and wind drift, and most that did were too water-logged or, like the sea heart, too far from home to root and grow unaided.

A few came from America, no one really knows by what device. It is thought that seeds of some orchids native to North America came airborne to Ireland. Sea peas have drifted over from the Great Lakes or the northeast coast of the United States. But what brought the American pipewort? Close by Dog's Bay, Nelson led me into the mist over low hills with pools in their hollows. There, in one pool, grew unimposing stems with primitive heads poking above the surface, a few inches of miracle. Pipewort does not belong here, yet it grows here.

We went on to the peat, the deep strata of aseptic, quaking sphagnum; to heathers and heaths, some of them growing here and in northern Spain and nowhere else; to the stumps of pines that had lived in these lands thousands of years ago in a different climate, preserved in the immaculate bog; and south of the bay to The Burren, the real hollow hills, great bald loaves of limestone ridged and runneled at the surface, a mass of caverns and tunnels carrying fresh water below. Some of these streams spew into the Atlantic from mouths in the rock hundreds of feet down. Nelson told me about walking the high limestone pavement, stopping to inspect plants growing in the clefts out of the wind, of walking past stone forts built a thousand years before a red-haired Genoese stepped ashore, dodging through the confusion of sherry casks and crates of the produce of Spain to walk the streets of Galway. As I listened, I kept seeing the chart of the North Atlantic I have studied for so long. I saw Ireland's northern tip parallel with Moscow, the whole island lying north of the major cities of the United States and Canada. Roses at Christmas? Orchids? What secret furnaces are at work here?

The man on the London double-decker will tell you it's the ruddy Gulf Stream at work. Ask coastal dwellers from France to Norway, and they'll tell you the Stream flows right past their beaches. The travel sections of the best newspapers say so. When *Welcome* arrived in Madeira, I bought a copy of the *International*

Herald Tribune, and there was a story about Jersey and Guernsey. "The Gulf Stream," read the subhead, "gives the [Channel] Islands a balmy climate even in spring and autumn."

The same conviction flourishes in America. Oceanographers at Woods Hole tell the story of a manufacturer who approached Columbus Iselin with a problem. He made perfume, he said, and he wanted to sell it in Europe. He was going to charter a tanker and dump his product off Florida. His aim was to have it in the nostrils of the British by Easter. When should he launch this seaborne promotion? Supporters of a project that would submerge huge turbines in the Florida Current to produce electricity (their blades would turn slowly enough so that whales could swim through unscratched) wanted to know if a big array of the machines would constrain transport enough to cool Europe's climate. And during one of our wars, some inspired souls came up with the idea of building a great dam out on the Grand Banks to deflect the Stream and freeze the Germans.

Henry Stommel once told me that a European can say his winter weather is a beneficiary of the Gulf Stream in the same way that someone getting his feet wet in a gutter blocks from a fire can say the fire hose was responsible. We still don't know the Stream's track much beyond the Grand Banks, though there are better guesses being made now. Everyone agrees that the Stream carries warm water in abundance to the high latitudes. It is also clear that anomalously warm water is found in winter off Norway, Ireland, most of the rest of coastal north Europe. But the connections are not all that obvious.

Nelson Hogg at the Woods Hole Oceanographic Institution studies transport in the Stream. He and his colleagues have measured it at the rate of around two hundred million cubic meters per second near the New England Seamounts, but only about thirty million cubic meters per second continue east in the general direction of Europe. The rest, Hogg says, recirculates, driven by eddies on both sides. The eddies don't actually entrain the water, carry it back upstream, and dump it in the current again. They are not symmetrical, so as they swirl, their eccentricities serve to induce the water nearby to flow in slow and massive recirculation.

The eddies and meanders are, in Hogg's view and that of many other oceanographers, processes that enable the Stream to stop being a stream, to let it re-enter the general circulation of the North Atlantic. All fluid on a rotating body possesses a certain vorticity. You can see a form of vorticity in a river by watching a floating branch turn in the current, acted upon by the shear created by different parts of the flow moving at different velocities. Water entering the Florida Current spins in a certain way relative to the earth's rotation, and that spin is slower than that of the water in higher latitudes. Hogg thinks that after the Stream leaves Hatteras, it begins to reconcile itself with the North Atlantic by writhing and shedding, rubbing hard against its walls (hardest against the northern, or inshore, one), to produce vorticities compatible with those of its host waters.

At no time has it been conclusively proved that the Gulf Stream itself continues on to Europe. Ben Franklin and many another thermometrist thought it did, but they were simply measuring warm surface water and wishfully transforming their measurements into currents. James Rennell in the nineteenth century and Val Worthington in the twentieth argued, for different reasons, that it didn't. Gunter Dietrich, at the University of Kiel, developed state-of-the-art charts of currents in the eastern North Atlantic. But Dietrich's successor at Kiel, a British oceanographer named John Woods, looked closely at the information from which the maps were drawn and found June data lumped in with winter data, numbers from different cruises set cheek by jowl. Dietrich did wonderfully well in patching things together, Woods told me, "but I think it would be unrealistic to pretend that it's what we take as a true statement" of the circulation. Woods's work had indicated that near the Grand Banks both the North Atlantic Current, moving off to the northeast, and the Azores Current, flowing to the southeast, may well not be the continuations of the Stream that conventional wisdom takes them to be.

Woods, who is now spending most of his time running the marine-science end of Britain's National Environment Research Council, has concentrated his studies on the effects of seasonal cycles on the upper one hundred to two hundred meters of the ocean. The physics, he says, is enormously complex. But he and his students at Kiel have produced an atlas and papers describing seasonal stratification of surface waters. Density layers outcrop at the surface of the North Atlantic and move two thousand kilometers and more to the north with the waxing of the heating season, returning south with the wane. Eddies in the layers tease jets of warm water across the ocean. The seasonal system, which interacts with the gyre circulation in unknown ways, can be altered by bad weather, particularly in winter and spring, making each year unique.

Woods showed me one map on which a band of closely packed lines stood out. "That's actually an extension of the Gulf Stream," he said. "According to Stommel's theory, which doesn't include this stratification, there is no way that that current goes all the way to Africa. But if you put in these seasonal effects of stratification, suddenly you find — amazingly — that there is this great line across here." That line and the warm current it implies are much too far south to affect European climate in any direct way, yet in searching farther north, Woods says he has found another line running from Georges Bank to Scotland under the tracks of storms that control Europe's climate. It is possible, he thinks, that this northern line "may provide the link with the Gulf Stream that climatologists have sought for two hundred years."

A new version of an old notion is getting some attention these days. One of its chief proponents is Wallace Broecker of the Lamont-Doherty Geological Observatory, north of New York City. Broecker is a geochemist interested in the influence of different water masses on oceanic circulation. His constructs are more general, his flows more diffuse, than those of the dynamicists. He pretty much ignores the Gulf Stream gyre in the North Atlantic; instead, he favors the idea of what he calls a conveyor belt, the northern end of which lies near Iceland. Cold winds from Canada blow across this northern end, cooling the water so that it sinks. It then travels south, as Stommel proposed, rounds the Cape of Good Hope, and enters the Indian and Pacific Ocean, where it warms and returns in a shallower and less salty movement to the Atlantic, flowing toward Iceland. The returning water runs slowly, for the most part in or near the main thermocline, perhaps eight hundred meters down. Some as yet unknown process moves it to the surface near Iceland. The winds cool it from, say, 10 to 2 degrees centigrade. That heat, removed by the westerlies, is what warms Europe.

It is the business of science to deal harshly with the untested, worrying at it with contrasting data and theories until it fits, or doesn't. Broecker's way of warming Europe is getting that kind of treatment. Some dynamicists say the conveyor belt construct is interesting but too simple, that the sources of relatively warm water off Ireland and other spots in the eastern North Atlantic are many and complex. Most scientists working with climate agree that winds coming off the North American continent do act, when properly freighted with moisture, to affect Europe's winters, but many think the action begins far west of Iceland. Some of the greatest heat loss in the North Atlantic, they say, occurs in the waters off the northern Maritimes. Researchers working in winter off Newfoundland have reported seeing the sea smoke as cold winds bellied across warm waters. The water there gives off heat to the atmosphere at the rate of almost one thousand watts per square meter. That is the equivalent of a perfectly decent power station set down in a square kilometer of ocean.

The blessed heat capacity of the ocean is what is at work here. The sea receives and stores what the atmosphere passes through, and the great currents carry that received heat to cooler regions and release it. If the Stream got no farther than the Tail of the Banks, its presence there would be a blessing to the lands out east. Cold, dry winter winds moving over it would become warmer, wetter winds, delivering heat to Europe in the air itself and in the rain the maritime storms deliver.

A British meteorologist, T. N. Palmer, wrote recently that studies by his colleagues had demonstrated a tantalizing connection between the swings in sea-surface temperature southeast of Newfoundland — close by where the International Ice Patrol flies — and European climate. It was found that when the surface temperature there is abnormally high in winter, so is rainfall over northern Britain. Cold anomalies may be related to atmospheric blocking over the continent. No one knows the cause of Gulf Stream thermal variability near Newfoundland, Palmer says. He thinks the answer lies in the influence of the atmosphere on the ocean; an easterly wind component applied to a model of the region produces a warming effect.

How strange to learn, then, that the blue god, the Heat Express, may have helped form the northern ice sheets of the glacial period. When North America and Africa parted company something under two hundred million years ago, a small and sluggish sea formed in the rift. By a hundred million years ago, the infant Atlantic had developed a shallow, clockwise gyre. The continents continued to creep apart, and until quite recently there were two basic components of the circulation. One was a globe-circling current running west through the Straits of Panama, into the Tethys, the sea that completely split the land masses east of the present Mediterranean, and on into the Atlantic again. The second was a proto-Stream, a current that at first may have followed a path northward outside the Antilles. Later, sedimentary evidence indicates, a strait formed between the Gulf of Mexico and the Atlantic, the Suwanee Channel, cutting across the top of the Florida peninsula. That channel may have accounted for most of the flow in the southern portion of the Stream until a few million years ago, when the action shifted to the Straits of Florida.

This early Stream was relatively weak, hugging the North American shore as it flowed into the newly developed Labrador Sea. From there, it is possible that some of the flow found its way to the Pacific, making a paleo-reality of that grail of European explorers, the Northwest Passage. Today, a small amount of Pacific water flows the other way, from the Pacific through the Arctic to the North Atlantic.

Fifty or sixty million years ago, the Atlantic converted from a tranquil sea to a real ocean. The separation of Greenland and Scandinavia and related events opened the Atlantic to the Arctic. That meant that, in time, thermal gradients developed north to south and top to bottom. Water in the north cooled and sank, and abyssal currents began to scour the sediments. "Commotion in the ocean had set in," wrote Charles Hollister and William Berggren, respectively a geologist and paleontologist at Woods Hole, "and continues with us today."

About three and a half million years ago, the Panamanian straits closed as the Isthmus rose from the sea. Hollister and Berggren think that the Atlantic water thus trapped could have added substantially to the flow of the Gulf Stream. The strengthened current would have carried much more warm water northward, furnishing the moisture for arctic winds to bear away, freeze, and deposit in drifts that the summer seasons could not melt. Three million years ago or so, the Arctic ice cap began a growth period. The Labrador Current became a respectable flow about then and forced the Stream south to its present range, below 45 degrees north. In the ensuing ice ages, the Stream moved its course laterally, swinging offshore when sea levels were high and back toward the coasts during lowstands.

A good deal of the evidence underlying hypotheses of paleoclimates comes from the study of ocean sediments, in particular the remains of tiny organisms that were around at the time. A look at the ratios of certain isotopes of oxygen can yield a rough picture of changes in ocean surface properties as the eras rolled. The bugs were instrumental in signaling what happened to the Stream during the last glaciation, which peaked eighteen thousand years ago. John Imbrie, of Brown University, one of the leading students of the ice ages, has written a popular book on the subject with his daughter Katherine. The Imbries describe the 100,000-year cycles of the major onsets of ice during the Pleistocene. "As the ice sheets expanded and contracted, and forests and prairies swept back and forth across Europe and Asia," they wrote, "the Gulf Stream swung back and forth like a gate hinged on Cape Hatteras." On the warm side of the cycles, the Stream extensions flowed close to what we accept as their normal courses, generally toward Britain. As the ice advanced again, the currents headed straight across toward Spain. John Imbrie works with a project based at the Lamont-Doherty Geological Observatory that has been studying climate worldwide. Its reports indicate that at the height of the last glacial advance, the Stream ran from 2 to 4 degrees centigrade cooler than at present, but that its system carried tropical species farther east than it does today.

All through recorded history we have been living in the cycles of the ice. The climate five thousand years ago was warmer than it is now. Five hundred years ago we were in a little ice age. A thousand years ago, the Norse were able to grow cereal grain near their settlements on Greenland, but when Christopher Columbus sailed west for the first time, Pope Alexander VI wrote that for eighty years or so no bishop or priest had been in residence at Garda, "situated at the ends of the Earth in Greenland." And that "shipping to that country is very infrequent because of the extensive freezing of the waters." As the little ice age settled in, the Greenland colonies, strangled by ice, perished.

The terrible climate of those times killed more people in some places than the Black Death did. Life was miserable in eastern Europe. Hubert Lamb, one of England's most respected climatologists, wrote that whole harvests were lost, often year after year, and that "the poor were reduced to eating dogs, cats, even children." In Scotland they still remember the "dear years," when famine was the way of life. At sea, the ice turned what had been rough but routine sailing into impossibility. Waters passable to the northmen were, only a few centuries later, lethal to so many of the English searching for the Northwest Passage.

Modern times are, to date, somewhat gentler, but we, particularly those of us living within the Maginots of industrial societies, have tended to ignore even the most obvious signals of change — the crop failures in the Soviet Union, the desertification of Africa, the catastrophic storms and fisheries failures that El Niño brings, and now the terrible drought in the grain lands of the United States. We scoff at the heavy clothing of past centuries, forgetting what the weather was like then. We do not remember that a cooling in Europe brought an end to the risqué costumes women had adopted after the French Revolution. The British have lived to regret their practice, initiated in warmer times, of installing plumbing on the outside walls of their houses. The current warming in Washington, D.C., has landscape gardeners warning their clients that some of the old standbys simply won't do anymore.

But worldwide warming is at last catching public attention, possibly because the public is helping to create it. Man the isolated hunter didn't do much to influence climate, beyond setting a few fires here and there. Man the farmer, the factory hand, automotive man in his teeming millions, has at last progressed to the point where he can. Only a century ago, for instance, the carbon cycle of the world obeyed natural rhythms in which plants of sea and land took carbon dioxide from the air in life and returned it in death. The removal of much of that biomass through burial in bogs and ocean sediments kept things pretty much in balance. But when man disinterred that buried plant material — coal, oil — and burned it, the amount of atmospheric carbon dioxide began to increase. Similarly, when manure on the fields was replaced with petrochemical fertilizers, nitrous oxides rose into the air. When more and more land went into feedlots and rice paddies, atmospheric methane increased. Natural sources continue to supply the air, of course. (It is a satisfaction to me that not a little atmospheric methane comes from the flatulence of countless termites.) The combined supply of the gases is growing in such a way that many investigators believe the world will warm by several degrees within the next century.

Carbon dioxide, methane, water vapor, and other substances are the so-called greenhouse gases which act to hold atmospheric

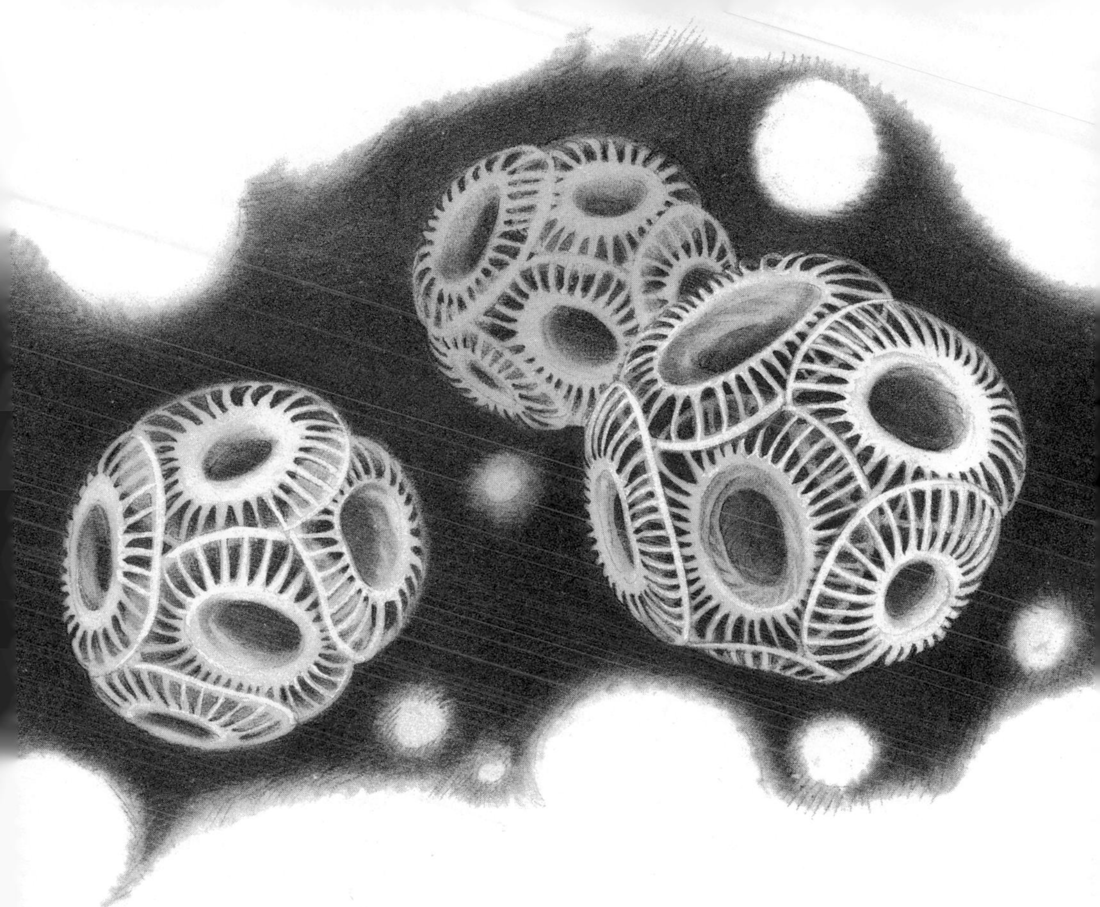

heat that would normally escape into space. Chlorofluorocarbons, used in refrigerators, insulating foams, and until recently in spray cans, also appear to attack the ozone layer high in the atmosphere, permitting more ultraviolet light to penetrate to earth and ocean and thus increasing risks of skin cancer in humans and possible species extinctions in marine plankton.

Warming by itself doesn't appear all that bad. Many areas of the world could stand a little. But wholesale warming means changes in climate. Rain belts will migrate, and what is now breadbasket may turn to dust. A warming sea means a rising sea. If the ocean were to heat throughout by just 1 degree centigrade, it would expand enough to raise sea level by about a foot and a half. Melting of terrestrial polar ice would add several feet to that.

A sensible approach to the future is through the past, and that is why studies of past climates take on such importance. The ocean, as ever, is one of the most promising laboratories for the work. Its enormous heat capacity, its ability to carry that heat from equator to pole and replace it with deep flows of cold, salty water, is central to establishing and changing climate. And since the Gulf Stream and some of its distributaries constitute the most powerful single engine of northward transport, they often come under the lens.

In the last glaciation, the North Atlantic Current was not able to sustain its role of drawing heat north. According to some theoreticians, that diminution in turn was due to a decrease in the formation of deep water north of Iceland. Perhaps it was the ice. So much water went into the glaciers and caps that sea level dropped two hundred or three hundred feet for a time. In addition, ice is a great insulator, and it could have been that the water under it simply never got cold enough to sink. In either case, less bottom water would have flowed out into the Atlantic and consequently less of an inflow from more southerly waters would have been required to replace it. The breakdown, rather sudden in earth terms, might happen again, for different causes. It is conceivable that the sea could warm so that water in high latitudes would not cool and sink in sufficient quantity to maintain "normal" ocean circulation.

Wallace Broecker believes that not enough warm water brought up the North American coast by the Gulf Stream can force its way across the winter ocean to warm Europe. That function is served, in his mind, by the conveyor belt running from Iceland around Africa and back. Not only does that system stoke Europe, it makes more bottom water in the process. He has applied this concept to the study of a cooling period that took place eleven thousand years ago, when forests that had come in as the glaciers retreated were ruined in decades. For a thousand years, shrubs grew instead of trees, and then things hopped back on the warming track.

Broecker thinks that a shrinkage in bottom water formation

was also responsible for this sudden shift, but the decrease this time was due to a redirection of meltwater from the great North American ice sheet. As the sheet pulled back, fresh water began to flow in enormous quantities down the St. Lawrence River and out into the North Atlantic, forming a cap of fresh water over the salt — fresh water too light to sink even when cooled. In winter it froze, forming an ice blanket against further cooling.

An underlying cause of ice ages, perhaps the major one, is thought to be slight shifts in the earth's orbit brought about by changes in gravitational attractions. The fluctuations produce changes in summer sunlight received at the earth's surface, and these produce expansions and contractions of ice caps with some regularity. But the process Broecker and many others have described is irregular, something like frequency control in a car radio: the control holds on to a station as long as it can and then jumps away. Scientists drilling cores in the Greenland ice cap have analyzed pockets of air trapped at various levels. They have found that the concentrations of atmospheric carbon dioxide varied appreciably over time, that the shifts seemed to coincide with changes in air temperature over the ice, and that these shifts were sudden — matters of just a couple of centuries.

I have been fascinated by an idea advanced fairly recently by a British scientist named James Lovelock. It is called the Gaia hypothesis, and it holds that the planet in all its parts — land, sea, air, plants, and animals — is a self-regulating organism. If Lovelock is right, then the oceans are the blood of the world. They help to control temperature, and they deal with chemical excesses and deficiencies. In the North Atlantic, warm surface waters pick up carbon dioxide and other gases from the air as they move north. As the carbon dioxide dissolves, a good deal is taken up by phytoplankton through photosynthesis and is converted into various carbonate structures within the plants. When they die, in the sea or in the guts of predators, their remains sink into the abyss. Meanwhile, the rest of the dissolved gas continues north to sink with the cooling water. That sinking, together with the rain of skeletal material and fecal pellets from the surface, serves to remove carbon dioxide from circulation for decades and centuries. When the deep water, moving south, rises toward the surface of the tropical sea, the gases it carries escape again to the atmosphere.

I like to think of Hubert Lamb in the swirl of the greenhouse controversy. He sees this possible warming period as only a bump on a longer-term decline of temperatures heading toward more glaciation in another several thousand years. When we run out of fossil fuel, he thinks, we will eventually return to that downslope. Whether or not that proves to be true, oceanographers are gearing up for research, on scales never before attempted, to see how their medium affects global change.

A leading figure among these researchers is Carl Wunsch, of MIT. Before I went to Ireland, I had been talking with Wunsch about acoustic tomography and about his plans for global study. But the excitement of the flowers and bogs put him out of my mind, until I stood in the mist with Charles Nelson, looking down at American pipewort growing in the pool. Not that Wunsch looks like an American pipewort. He is a slender man with a broad face and a voice that has some of the bite of the late pianist Oscar Levant, Gershwin's friend and general curmudgeon. I know of no other oceanographer so willing to discuss the blemishes of his science or so witty in the discussion.

The Wunschian idea that has stuck longest with me is that the ocean is largely unobserved. It is a point he never tires of making. He got up in front of a navy audience recently with visual aids depicting the number of oceanographic instruments deployed in the Pacific in 1981 and the number of meteorological instruments deployed in Europe in 1653 — "Actually an amazing number of barometers operating then." The Europeans, of course, won.

I learned from Wunsch that a fluid is by definition indivisible, an entity not satisfactorily investigable piece by piece. He gave me as example Val Worthington's report that after a cold winter the Gulf Stream got stronger. How do you actually show that? he asked. If it did, was it just a local thing or did the whole thing go faster? If the Stream carried a lot more water, where the hell did it go? Did it have any consequences six months later somewhere else? If it did, how would you find out what they were? "It's a bit like studying a beach with a magnifying glass," he said. "You may learn a lot about sand on the local scale, but you have no idea what formed that beach or how it is changing."

The subject of global change is a lot more important now than it was when I began this book. Each new look at the ozone deficiency, for example, feeds a new kind of collective anxiety, the realization that comeuppance is forever stalking exploitation. Scientists are making governments nervous with their warnings, and it is easier now to put global research on a properly international basis. The World Climate Research Program is such an umbrella, its mandate to determine human influence on climate and to work toward climate predictions of months, years, and decades. That last time scale is where Carl Wunsch — and John Woods, and John Gould, and a number of other scientists I've been describing — will be working. Their program is called the World Ocean Circulation Experiment.

WOCE, which will begin intensive field operations in 1991, is a child of the current technological revolutions in oceanography and remote sensing. It will map circulation and circulation changes using the latest versions of the European batfish, floats like Tom Rossby's RAFOS, and instruments that, like the tomographic transceiver, are still running through their tests. The last time I was in Woods Hole, a friend told me he was working on a pop-up device for WOCE, a profiler that went down close to the bottom, reversed, and on the return homed in on a sonic buoy deployed on the surface. No more searching for some electronic waif drifting and beeping in the ruck. Fun, said my friend, beaming.

WOCE will make use of several satellites. The jewel among them is the French-American Topex-Poseidon, which will carry the most precise oceanographic altimeter flown to date. Ships, buoys, and satellites will produce an ungodly amount of data over the decade of this massive program, much more than to-

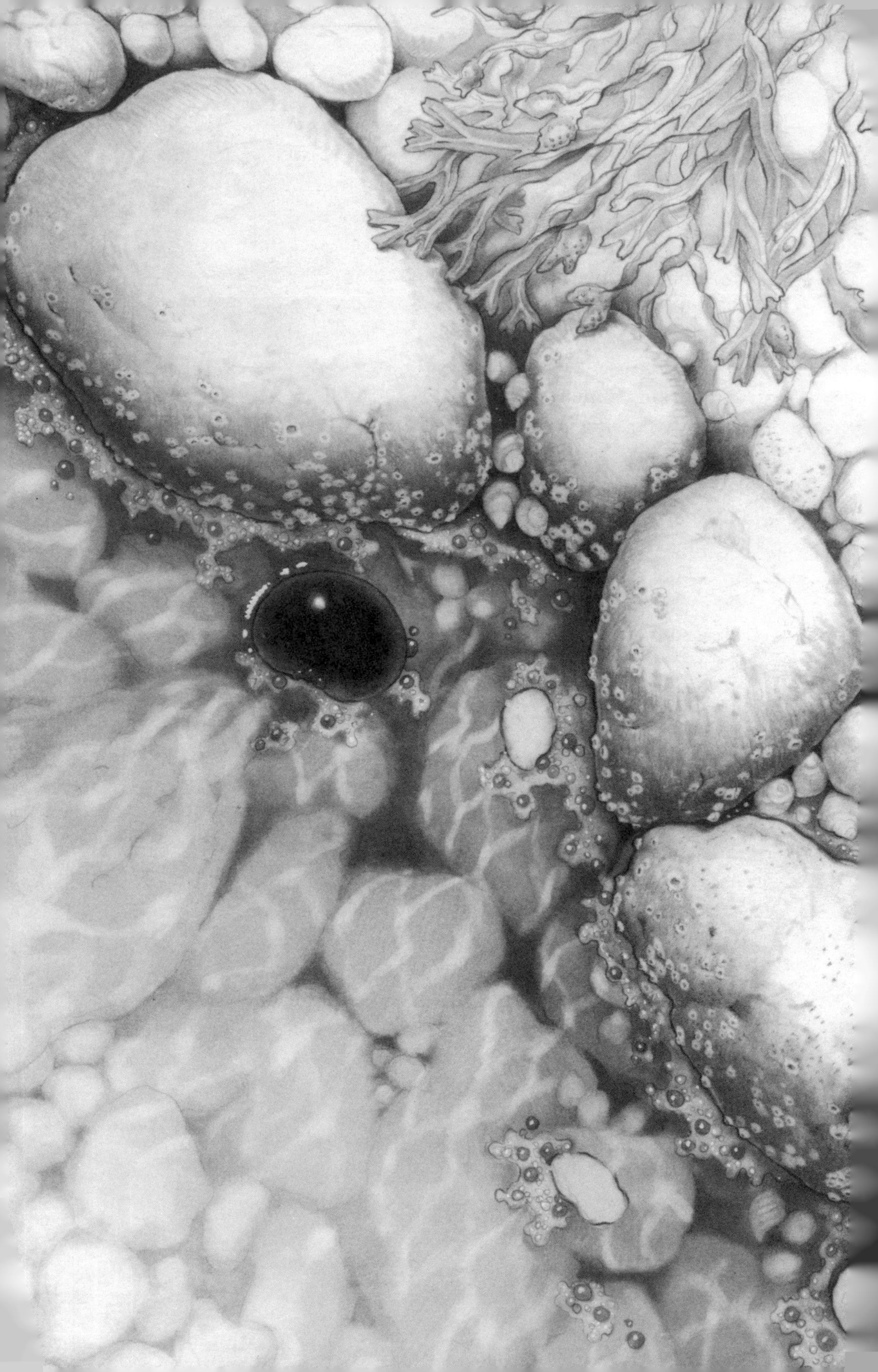

day's computers can handle. But John Woods told me that the evolution of computer power is itself almost predictable: the machines should be ready when WOCE is. Programmers are now under heavy pressure to develop software for the machines that can manage the oceans of information anticipated.

Modeling is moving along, and WOCE is counting on that movement. The program's architects believe that computers to come will enable them by the end of the century to develop models of global ocean circulation that can resolve eddies, those essential gears of the marine engine, and other features that are becoming increasingly important in the understanding of ocean dynamics. These simulations will be generations beyond the essentially laminar, nonlinear models that have shaped current knowledge. But even they, and the massive data sets on which they feed, will have to be stretched to accomplish the larger goal: merger with atmospheric models to make coupled representations for climate forecasters.

Oceanography has its divisions, though they seem not to be as entrenched as those in other mediums. Yet eddies of a sort are at work in the science, moving things around. Biologists are learning about physical oceanography, and chemists and physicists are learning about sea life. Impossible words like biogeochemistry swim at the surface. Biogeochemistry means that currents carry nutrients that feed populations whose ebb and flow can affect the uptake and disposition of gases whose concentrations affect climate, changes in which can alter currents. Consider the coccolithophore, a tiny organism whose calcareous skeletal plates, enlarged by electron microscopy, look like the mountains of the moon. Massed together in bloom, the creatures gleam brightly in the images produced by satellite color scanners. Coccolithophores are one of two groups of plankton most active in removing dissolved carbon dioxide from seawater. They also produce a gas called dimethylsulfide, which when released to the atmosphere provides an important source of nuclei for the formation of water droplets and therefore of clouds. Changes in their populations could thus result in changes in cloud cover and, because of the carbon dioxide connection, changes in the global greenhouse.

We know very little. Many an accepted explanation of oceanic workings has never been demonstrated. We have no theory explaining the dynamics of the Gulf Stream east of Hatteras, says Carl Wunsch. The ocean is still largely unobserved.

And so to Galway. I walked Dog's Bay Beach, where Charles Nelson finds sea hearts and other visitors from the west after equinoctial storms. I wondered how they get here, by what combination of wind and drift, and if that combination resembles what we now have in the way of circulation maps. I thought of the fuchsia escaped from the walled gardens around me, the gardens growing a thousand years. I set that against the impatience of other cultures. In them — in my own — we cannot wait for nature. We fill in wetlands, plow and pave where we shouldn't, proclaiming God-given dominion. Long ago, a British crusader sang:

> O western wind, when wilt thou blow?
> That the small rain down can rain?
> Christ, that my love were in my arms,
> And I in my bed again.

The wind he wanted was the southwest wind, the wind of rain and promise, Europe's good wind, the beam wind that would take him home. Today, in our *Tampa*s, we go when we will, deferring only to the tempest. But I think we are coming to appreciate the price we pay for impatience.

Columbus was here, nearby, in Galway. They say that here, or in Spain, he found a pilot for his first voyage, a Galwayman named Rice de Galvey or Penrise, and that the man had previously made voyages to the American continent. They say Columbus worshiped at a church in the center of the city, now called the Collegiate Church of St. Nicolas, that was built by Anglo-Normans in 1320. For the last time, I conjured up this red-haired man, walking the city, thinking of sea beans and Cipangu. We can answer his questions far better than he could, but we are very much where Columbus was in trying to answer our own questions. He addressed himself to the location of Cathay; we must address ourselves to the preservation of a planet. Our answers, like his, will be right. And like his, they will be wrong. Perhaps like him, we will be lucky in the voyage.

11
The Cathedral

I HAVE SEEN wind against current, and the sharp waves. I have looked through borrowed binoculars at a line far off on the sea and been told it was the west wall. I have remarked on a sudden change of water color and learned it was not the Stream but the termination of coastal water. I have studied what I could of plunging isothermals and tried to bring them to an image of understanding. But when I came to write this coda, I was not at all sure that in three years of trying I had ever truly sensed the blue god.

Aorta once seemed a good metaphor. It had arisen in conversations with a friend, a retired oceanographer living in a snug house of his own design on a hill near Woods Hole. The retired oceanographer spent time educating me in the way the ocean has of moving heat to the poles. The mechanism, which he likened in passing to arterial flow, is, he said, "probably the most vital thing we have to understand about the oceans and atmosphere, because it determines the habitability of the earth."

The problem with the aorta, said a theoretician of my acquaintance, is that it goes too far; it carries vivid images of its own that are extraneous to the problem: "The aorta conveys the idea of a tube. Yet in the Gulf Stream, there is no tube; the flow makes its own tube." Also, he said, the Stream is involved with more than the transport of fluid. In its passage, it must satisfy the demands of certain balances, like momentum and energy.

All right, all right! I said. Then how do I make out this monstrous current? You scientists use numbers and equations and models. Is that reality? What, asked the theoretician, do you mean by reality? We're interested in things like how the ocean works dynamically. People with that interest go to sea, puke their guts out, deploy their instruments, and come back with numbers from which they try to infer something. Is that the real ocean? he asked. Or is it, like the aorta, a construct of the mind?

My oceanographer friend would certainly agree with the theoretician, especially about constructs. Since retiring, he has given himself over to the most varied interests, writing haiku, experimenting with novel ways to heat his house, considering the abundances and limitations of the world. He is, I think, a natural philosopher, as scientists were called before they knew who they were. He considers himself a Cartesian, and he works at learning how we invent ourselves and our world "with three pounds of meat at the top of our necks."

"Science is not the key to the Gulf Stream," my oceanographer friend said, sitting in his snug house. His glasses gave his eyes a wondering look at odds with his gyrene haircut. He sucked on a cigarette. "Science has a way of talking about these things, but the key is human understanding of the nature of the physical world. So keep philosophy in mind." He dug out his Bacon and found what he wanted. "Here! They should have this carved into the lintels of every office in Washington: 'Nature, to be commanded, must be obeyed.' " I took the book and read further, with some despair, for Bacon was asking the impossible of me: "Those, therefore, who determine not to conjecture and guess but to find out and know, not to invent fables and romances of words, but to look into . . . the nature of this real world, must consult only things themselves."

What things? I have been dealing with a being co-existing with itself on innumerable scales, this god of the great flow, of its snaking, eddy-shedding return to the ocean that spawned it, of its coupling with the winds that temper a continent. Consult what things? Here, at the end of my work, the facts I gathered are fading. The brightest things of my memory are fables and romances — and the vision of one dive.

There is a cliff that steps down through a mile of Bahamian water flowing into the Florida Current. When Larry Madin and his crew had finished their work, we went there. The dive boat was full that morning. Everyone wanted to sport in the warm sea, unencumbered by tethers, before heading north. We chose partners and rolled backward into the water.

Two bottoms lay below, sand at a hundred feet and, beyond, darkling blues that shut out the immodesty of the black under them, the cold, the heavy pressures. As we sank, we felt the currents nudging down over the lip. Young dorados, wrasses, grunts, unnameable others turned and flashed in their squadrons. Eagle rays patrolled, barracuda, a hawkbill turtle. A large shark idled under a formation, too far away, too indolent, to raise the heartbeat.

We sank below the lip, and I forced myself to let go of the sight of coral and hunting fish, to look away and down. I felt the motherly hug of deep water. The current pressed slightly on my back. I tipped a little air into my buoyancy vest to keep level, and that small act of self-maintenance generated thoughts of divers who let themselves go, drunk on nitrogen or sober and looking for death. A decade before his suicide by shotgun, Ernest Hemingway lay in water like this, but right in the Stream. It was awfully nice down there, he said later, and he was tempted to stay.

It *was* awfully nice. Perhaps I had a little touch of narcosis: a piano sounded in my mind. I made out the piece, one I used to play when I was young and considered myself a pianist — Debussy's "La Cathédrale Engloutie." That seemed logical, Mr. Bacon. Those divers drifting over the abyss could be supplicants. Those bubbles, lofting, tumbling toward the surface, could be columns. The colors were holy enough. This could be a sunken cathedral. This, for me, could be the way in to the blue god.

Selected Bibliography

WITH FEW EXCEPTIONS, the books listed below represent works that are either in print or probably can be located without excessive searching. Except as noted, they are books accessible to the reader interested in pursuing matters relating to the Gulf Stream in the lay literature. Those interested in oceanographic source material might contact the public affairs offices of such marine-science centers as the Woods Hole Oceanographic Institution, the Graduate School of Oceanography at the University of Rhode Island, or the Rosenstiel School of Marine and Atmospheric Science at the University of Miami.

Boorstin, Daniel J. *The Discoverers.* New York: Random House, 1983.

Carson, Rachel L. *The Edge of the Sea.* Boston: Houghton Mifflin, 1979.

———. *The Sea Around Us.* New York: Oxford University Press, 1951.

———. *Under the Sea Wind.* New York: Mentor Books, 1955.

Couper, Alistair, ed. *The Times Atlas of the Oceans.* New York: Van Nostrand Reinhold, 1983.

Crosby, Alfred W. *Ecological Imperialism: The Biological Expansion of Europe, 900–1900.* New York: Cambridge University Press, 1986.

Daley, Robert. *Treasure.* New York: Pocket Books, 1986.

Deacon, Margaret. *Scientists and the Sea: A Study of Marine Science.* London: Academic Press, 1971.

Gleick, James. *Chaos: Making a New Science.* New York: Viking Penguin, 1987.

Gross, Grant M. *Oceanography: A View of the Earth.* 4th ed. Englewood Cliffs, N.J.: Prentice-Hall, 1987.

Hemingway, Ernest. *Islands in the Stream.* New York: Charles Scribner's Sons, 1970.

Houghton, John T. *The Global Climate.* Cambridge: Cambridge University Press, 1984. Not for the general reader.

Imbrie, J.I., and Imbrie, K.P. *Ice Ages.* Hillside, N.J.: Enslow Publishers, 1979.

Lamb, H.H. *Climatic History and the Future.* Princeton, N.J.: Princeton University Press, 1985.

Meinig, D.W. *The Shaping of America: A Geographical Perspective on Five Hundred Years of History.* Vol. 1, Atlantic America, 1492–1900. New Haven: Yale University Press, 1986.

Morison, Samuel Eliot. *Admiral of the Ocean Sea: A Life of Christopher Columbus.* Boston: Northeastern University Press, 1983.

———. *The European Discovery of America: The Southern Voyages, A.D. 1492–1616.* New York: Oxford University Press, 1974; *The Northern Voyages, A.D. 500–1600.* New York: Oxford University Press, 1971.

National Geographic Society. *Atlas of the World.* Washington, D.C.: National Geographic Society, 1981.

O'Meara, J.J., trans. *The Voyage of Saint Brendan: "Journey to the Promised Land."* Atlantic Highlands, N.J.: Dolmen Press/Humanities Press, 1981.

Parry, J.H. *The Discovery of the Sea.* Berkeley and Los Angeles: University of California Press, 1981.

Pond, S., and Pickard, G.L. *Introductory Dynamical Oceanography.* 2d ed. New York: Pergamon Press, 1983. Not for the general reader.

Schlee, Susan. *The Edge of an Unfamiliar World: A History of Oceanography.* New York: E.P. Dutton, 1973.

———. *On Almost Any Wind: The Saga of the Oceanographic Research Vessel "Atlantis."* Ithaca, N.Y.: Cornell University Press, 1987.

Severin, Tim. *The Brendan Voyage.* London: Arrow Books, 1979.

Stewart, Robert H. *Methods of Satellite Oceanography.* Berkeley and Los Angeles: University of California Press, 1985.

Stommel, Henry. *The Gulf Stream: A Physical and Dynamical Description.* Berkeley and Los Angeles: University of California Press, 1976. Not for the general reader.

———. *A View of the Sea: A Discussion Between a Chief Engineer and an Oceanographer about the Machinery of the Ocean Circulation.* Princeton, N.J.: Princeton University Press, 1987. For the reader with technical background.

Teal, John, and Teal, Mildred. *The Sargasso Sea.* Boston: Atlantic Monthly Press, 1975.

Van Doren, Carl. *Benjamin Franklin.* New York: Viking Press, 1938.

Warren, B., and Wunsch, C., eds. *Evolution of Physical Oceanography.* Cambridge, Mass.: MIT Press, 1981.

Whipple, A.B.C. *Restless Oceans.* New York: Time-Life Books, Planet Earth Series, 1983.

Wilford, John Noble. *The Mapmakers: The Story of the Great Pioneers in Cartography, from Antiquity to the Space Age.* New York: Random House, 1982.

Index